INTRODUCTION TO
BUSINESS SYSTEMS ANALYSIS

Graham Curtis • Jeffrey A. Hoffer
Joey F. George • Joseph S. Valacich

U N I V E R S I T Y   O F   P H O E N I X
COLLEGE OF INFORMATION SYSTEMS AND TECHNOLOGY

Pearson
Custom
Publishing

This special edition published in cooperation with Pearson Custom Publishing

Printed in the United States of America

10 9 8 7 6 5

Please visit our website at www.pearsoncustom.com

ISBN 0–536–60253–0

BA 990637

PEARSON CUSTOM PUBLISHING
75 Arlington Street, Suite 300/Boston, MA 02116
A Pearson Education Company

CONTENTS

CONGRATULATIONS...

You have just purchased access to a valuable website that will open many doors for you! The University of Phoenix has chosen to enhance and expand your course's material with a dynamic website that contains an abundance of rich and valuable online resources specifically designed to help you achieve success!

This website provides you with material selected and added to powerful online tools that have been seamlessly integrated with this textbook, resulting in a dynamic, course-enhancing learning system. These exciting tools include:

> Online Study Guide
> Online glossary
> Links to selected, high-quality websites
> And more!

You can begin to access these tremendous resources immediately!

www.pearsoncustom.com/uop: The opening screen of the University of Phoenix website includes book covers of all the Pearson Custom Publishing books in the BSBIS and BSIT programs. Click on the book cover representing your course. This will launch the online study guide for the course in which you are currently enrolled **and** the glossary of key terms for all the University of Phoenix BSBIS and BSIT courses.

CD-ROM: The accompanying CD-ROM includes key terms underlined within the online book that are linked to the World Wide Web. Use the enclosed CD-ROM to launch websites selected to reinforce your learning experience.

TAKE THE FIRST STEP ON THE ROAD TO SUCCESS TODAY!

BUSINESS INFORMATION SYSTEM DEVELOPMENT

THE PARTICIPANTS IN ANALYSIS AND DESIGN

It is common for a computer systems project to be initiated because someone has recognized that a problem exists with the way that things are currently done. Alternatively, an opportunity is perceived that will lead to an improvement on the present system. In either case the users of the existing system will play an important role. They can provide information on the current system. They will also be able to specify, in their own terms, the requirements of the new system.

Programmers are responsible for turning those requirements into programs. When executed these will control the operation of the computer, making it perform to serve the needs of users. However, the programmer will be a computer specialist and will see the problem in computer terms. He or she will be talking a different language from the users. There is a communications gap.

The gap is filled by the systems analyst. This person is able to understand and communicate with users to establish their requirements. The analyst will also have an expert knowledge of computers. He or she will reframe those requirements in terms that programmers can understand. Code can then be written. It is important that the analyst is a good communicator and can think in terms of the user's point of view as well as that of the programmer (see Figure 1-1).

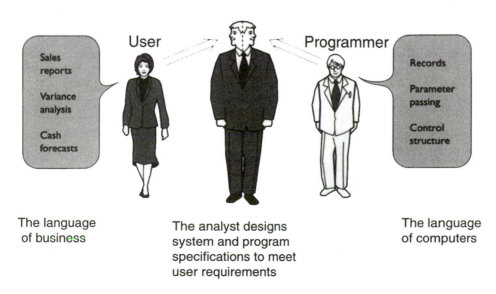

| The language of business | The analyst designs system and program specifications to meet user requirements | The language of computers |

FIGURE 1-1   *The Role of the Analyst*

This translation of requirements is not a straightforward process. It is not, for instance, like translation from German into English. It is more helpful to think of it along the lines of architecture and building. The client (user) states his or her understanding of what the building should look like and what functions it should perform (user statement of the requirements of the system). The architect (analyst) then takes these intentions and provides a general sketch of the building that will satisfy them (logical model of the intended system). Having agreed this with the client (user) the architect (analyst) then draws up a detailed blueprint of the design (detailed program specification) from which the builders (programmers) can work. Just as the architect's task is a skilled one requiring a knowledge of building materials so the analyst needs a knowledge of computers.

The analyst's task is not restricted to providing specifications for the programmers. The analyst has a range of responsibilities:

1. The analyst is responsible for investigating and analyzing the existing system as to its information use and requirements.

2. The analyst judges whether it is feasible to develop a computer system for the area.

3. The analyst designs the new system, specifying programs, hardware, data structures and control and other procedures.

4. The analyst will be responsible for testing and overseeing the installation of the new system, generating documentation governing its functioning and evaluating its performance.

The analyst is likely to come from either a computer science or a business background. He or she will usually possess a degree and/or be professionally qualified. Sometimes an analyst may have risen 'through the ranks' from programmer to programmer/analyst to a full systems analyst.

As well as possessing significant technical skills in computing, the analyst must fully appreciate the environment and work practices of the area within which the computer system will be used. Knowledge and experience are necessary but not sufficient. The analyst must, above all, be a good communicator with business personnel as well as with the technical staff. He or she must be able diplomatically to handle conflicts of interest that inevitably arise in the course of a project. Managerial, particularly project management, skills are another essential asset as a project involves a complex interaction of many people from different backgrounds working on tasks that all have to be coordinated to produce an end product. The design process is not mechanical and the analyst must demonstrate both considerable creativity and the ability to think laterally. Finally, analysts need to exude confidence and controlled enthusiasm. When things go wrong it will be the analyst to whom people look as the person to sort out the problems and smooth the way forward.

YOUR ROLE AND OTHER ORGANIZATIONAL RESPONSIBILITIES IN SYSTEMS DEVELOPMENT

In an organization that develops its own information systems internally, there are several types of jobs involved. In medium to large organizations, there is usually a separate Information Systems (IS) department. Depending on how the organization is set up, the IS department may be a relatively independent unit, reporting to the organization's top manager. Alternatively, the IS department may be part of another functional department, such as Finance, or there may even be an IS department in several major business units. In any of these cases, the manager of an IS department will be involved in systems development. If the department is large enough, there will be a separate division for systems development, which would be homebase for systems analysts, and another division for programming, where programmers would be based (see Figure 1-2). The people for whom the systems are designed are located in the functional departments and are referred to as users or *end-users*.

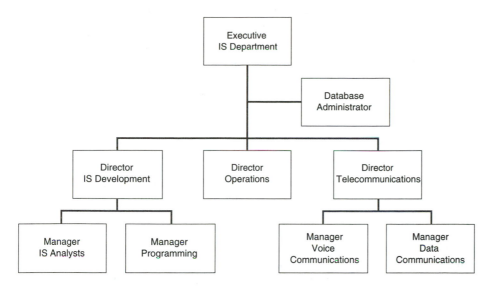

FIGURE 1-2   *Organizational Chart for Typical IS Department*

Some organizations use a different structure for their IS departments. Following this model, analysts are assigned and may report to functional departments. In this way, analysts learn more about the business they support. This approach is supposed to result in better systems, since the analyst becomes an expert in both systems development and the business area.

Regardless of how an organization structures its information systems department, systems development is a *team effort*. Systems analysts work together in a team, usually organized on a project basis. Team membership can be expanded to include IS managers, programmers, users, and other specialists who may be involved (throughout or at specific points) in the systems development project. It is rare to find an organizational information system project that involves only one person. Thus, learning how to work with others in teams is an important skill for any IS professional, and we will stress team skills throughout this book.

A good team has certain characteristics, some that are a result of how the group is assembled and others that must be acquired through effort on the part of team members (see Table 1.1). A good team is diverse and tolerant of diversity:

- A diverse team has representation from all the different groups interested in a system, and the representation of these groups on the team increases the likelihood of acceptance of the changes a new system will cause.

- Diversity exposes team members to new and different ideas, ideas they might never think of were all team members from the same background, with the same skills and goals.

- New and different ideas can help a team generate better solutions to its problems and defend the course of action it chooses.

TABLE 1.1
Characteristics of Successful Teams

- Diversity in backgrounds, skills, and goals
- Tolerance of diversity, uncertainty, ambiguity
- Clear and complete communication
- Trust
- Mutual respect and putting one's own views second to the team
- Reward structure that promotes shared responsibility and accountability

- Team members must be able to entertain new ideas without being overly critical, without dismissing new ideas out of hand simply because they are new.
- Team members must be able to deal with ambiguous information as well as with complexity and must learn to play a role on a team (and different roles on different teams) so that the talents of all team members can best be utilized.

In order to work well together, a good team must strive to communicate clearly and completely with its members. Team members will communicate more effectively if they trust each other. Trust, in turn, is built on mutual respect and an ability to place one's own goals and views secondary to the goals and views of the group. To help ensure that a team will work well together, management needs to develop a reward structure that promotes shared responsibility and accountability within the team. In addition to rewards for individual efforts, team members must be rewarded by IS managers for their work as members of an effective work unit.

Team success depends not only on how a team is assembled or the efforts of the group but also on the management of the team. Reward systems are one part of good team management. Effective project management is another key element of successful teams. Project management includes devising a feasible and realistic work plan and schedule, monitoring progress against this schedule, coordinating the project with its sponsors, allocating resources to the project, and sometimes even deciding whether and when a project should be terminated before completing the system.

The characteristics of each systems analysis and design project will dictate which types of individuals should be on the project team. In general, those involved in systems development include IS managers, systems analysts, programmers, end-users, and business managers as well as additional IS managers, technicians, and specialists. We will now preview the role of each of these players and other stakeholders in systems development.

■ IS MANAGERS IN SYSTEMS DEVELOPMENT

The manager of an IS department may have a direct role in the systems development process if the organization is small or if that is the manager's style. Typically, IS managers are more involved in allocating resources to and overseeing approved system development projects rather than in the actual development process. Thus, IS managers may attend some project review meetings and certainly will expect written status reports on project progress covering their areas of concern. IS managers may prescribe what methodologies, techniques, and tools are to be used and the procedure for reporting the status of projects. As department leaders, IS managers are also responsible for career planning and development for systems analysts and other employees and for solving problems that arise in the course of development projects.

There are, of course, several IS managers in any medium to large IS department (see Figure 1-2). The manager of an entire IS department may have the title Chief Information Officer and may report to the president or chairman of the firm. Each division of the IS department will also have a manager. Typical titles for these managers are Director of IS Development, IS Operations Manager, and IS Programming Director. The Director of IS Development may be responsible for several development projects at any given time, each of which has a project manager. The responsibilities and focus of any particular IS manager depend on his or her level in the department and on how the organization manages and supports the systems development process.

■ SYSTEMS ANALYSTS IN SYSTEMS DEVELOPMENT

Systems analysts are the key individuals in the systems development process. To succeed as a systems analyst, you will need to develop four skills: analytical, technical, managerial, and interpersonal. *Analytical skills* enable you to understand the organization and its functions, to identify opportunities and problems, and to analyze and solve problems. One of the most important analytical skills you can develop is systems thinking, or the ability to see organizations and information systems as systems. Systems thinking provides a framework from which to see the important relationships among information systems, the organizations they exist in, and the environment in which the organizations themselves exist. *Technical skills* help you understand the potential and the limitations of information technology. As an analyst, you must be able to envision an information system that will help users solve problems and that will guide the system's design and development. You must also be able to work with programming languages, various operating systems, and computer hardware platforms. *Management skills* help you manage projects, resources, risk, and change. *Interpersonal skills* help you work with end-users as well as with other analysts and programmers. As a systems analyst, you will play a major role as a liaison among users, programmers, and other systems professionals. Effective written and oral communication, including competence in leading meetings, interviewing, and listening, is a key skill analysts must master. Effective analysts successfully combine these four skills, as Figure 1-3, a typical advertisement for a systems analyst position, illustrates.

As with any profession, becoming a good systems analyst takes years of study and experience. Once hired by an organization, you will generally be trained in the development methodology used by the organization. There is usually a career path for systems analysts that allows them to gain experience and advance into project management and further IS or business management. Many academic IS departments train their undergraduate students to be systems analysts. As your career progresses, you may get the chance to become a manager inside or outside the IS area. In some organizations, you can opt to follow a technical career advancement ladder. As an analyst, you will become aware of a consistent set of professional practices, many of which are governed by a professional code of ethics, similar to other professions.

■ PROGRAMMERS IN SYSTEMS DEVELOPMENT

Programmers convert the system specifications given to them by the analysts into instructions the computer can understand. Writing a computer program is sometimes called writing code, or *coding*. Programmers also write program documentation and programs for testing systems.

SYSTEMS ANALYST
Distribution Center

We are the world's leading manufacturer of women's intimate apparel products. Our organization in the Far East has an opening for a Systems Analyst.

Requirements:

- Bachelor's degree (or comparable) in Computer Science and/or Business Administrations with 5 (+) years of working experience.

- In-depth understanding of Distribution and Manufacturing concepts (Allocation, Replenishment, Ship Floor Control, Production Scheduling, MRP)

- Working knowledge of project management and all phases of software development life cycle.

- Experience with CASE tools, PC and Bar Code equipment is essential.

- Working knowledge of an AS/400 and/or UNIX environment with the languages C, RPG400 and/or COBOL are desirable

The successful candidate will provide primary interface for all user problems, answer technical questions and requests within the applications development group; work with user areas to establish priorities; and provide recommendations and directions for process improvement through automation; skills in an Asian language is a plus.

We offer an attractive compensation package, relocation assistance and the technical and analytical challenges you would expect in a state-of-the-art environment. The position will report to Senior Management.

Please forward your resume, along with salary requirements to:

FIGURE 1-3   *Typical Job Ad for a Systems Analyst*

For many years, programming was considered an art. However, computer scientists found that code could be improved if it were structured, so they introduced what is now called *structured programming* (Bohm and Jacopini, 1966). In structured programming, all computing instructions can be represented through the use of three simple structures: sequence, repetition, and selection. Becoming a skilled programmer takes years of training and experience. Many computer information systems undergraduates begin work as programmers or programmer/analysts.

Programming is very labor-intensive, therefore, special-purpose computing tools called *code generators* have been developed to generate reasonably good code from specifications, saving an organization time and money. Code generators do not put programmers out of work; rather, these tools change the nature of programming. Where code generators are in use, programmers take the generated code and fix problems with it, optimize it, and integrate it with other parts of the system. The goal of some computer-aided software engineering (CASE) tools is to provide a variety of code generators that can automatically produce 90 percent or more of code directly from the system specifications normally given a programmer. When this goal is achieved, the role of programmers on systems development teams will be changed further.

■ END-USERS IN SYSTEMS DEVELOPMENT

In the typical traditional systems development process, systems analysts interview, survey, and observe end-users to determine what they need from an information system. End-users are business professionals who are experts in their fields but who usually do not have the skills, time, desire, or responsibility required to develop information systems. Therefore, as an analyst, you will work with users to convert their knowledge of the business into supportive information systems. End-users often are your clients or customers, the people for whom you are building a system. In many cases, end-users will also serve on the systems development team, providing their expertise in very active ways. You and other IS professionals will also provide support and assistance for those more sophisticated end-users who write, test, and implement their own information or data distribution systems.

SUPPORTING END-USER DEVELOPMENT   As the number of user requests for new or improved information systems increases, an IS department will have to assign priorities to development projects. Prioritization means that some users will get their systems right away while others must wait. In the 1970s, with the spread of time-sharing systems that enabled several people to use the same computer at the same time, it became technically possible for end-users to develop their own systems. As education and experience made end-users aware of these technologies and they became skillful in using them, many users, impatient for their requested projects to be scheduled, developed their own applications. Consequently, a significant role for systems analysts and other IS professionals is to help end-users develop their own systems, which are often stand-alone systems or data distribution systems designed to enhance an existing system developed by IS professionals.

Support of end-user development has several dimensions. First, IS professionals must evaluate and make available suitable tools for end-users. Second, IS professionals must train end-users in the use of these tools and their proper application. Third, IS professionals must be available to assist end-users and to answer questions or perform more complicated systems development work whenever end-users have difficulties. Finally, IS professionals must continue to maintain the data capture and transfer applications that manage the databases from which end-users extract the data needed in the systems they build.

End-user design and development of information systems has been somewhat controversial in the past (Alavi and Weiss, 1985; Davis, 1982). Some IS managers have worried about the quality of the systems end-users produce. However, some end-users believe there is no choice, especially if users must wait for the IS department to provide the desired and needed systems. Of course, if an organization is too small to have an IS department and outside consultants are too expensive, end-users have little choice but to develop their own systems or settle for off-the-shelf software.

■ Business Managers in Systems Development

Another group important to systems development efforts is business managers, such as functional department heads and corporate executives. These managers are important to systems development because they have the power to fund development projects and to allocate the resources necessary for the projects' success. Because of their decision-making authority and knowledge of the firm's lines of business, department heads and executives are also able to set general requirements and constraints for development projects. In larger companies where the relative importance of systems projects is determined by a steering committee, these executives have additional power as they are usually members of the steering committees or systems planning groups. Business managers, therefore, have the power to set the direction for systems development, to propose and approve projects, and to determine the relative importance of projects that have already been approved and assigned to other people in the organization.

■ Other IS Managers/Technicians in Systems Development

In larger organizations where IS roles are more differentiated, there may be several additional IS professionals involved in the systems development effort. A firm with an existing set of databases will most likely have a *database administrator* who is usually involved in any systems project affecting the firm's databases. Network and *telecommunications experts* help develop systems involving data and/or voice communication, either internal or external to the organization. Some organizations have *human factors* departments which are concerned with system interfaces and ease-of-use issues, training users, and writing user documentation and manuals. Overseeing much of the development effort, especially for large or sensitive systems, are an organization's *internal auditors* who ensure that required controls are built into the system. In many organizations, auditors also have responsibility for keeping track of changes in the system's design. The necessary interaction of all these individuals makes systems development very much a team effort.

INTERPERSONAL SKILLS

Although, as a systems analyst, you will be working in the technical area of designing and building computer-based information systems, you will also work extensively with all types of people. Perhaps the most important skills you will need to master are interpersonal. In this part of the section, we will discuss the various interpersonal skills necessary for successful systems analysis work: communication skills; working alone and with a team; facilitating groups; and managing expectations of users and managers.

■ Communication Skills

The single most important interpersonal skill for an analyst, as well as for any professional, is the ability to communicate clearly and effectively with others. Analysts should be able to successfully communicate with users, other information systems professionals, and management. Analysts must establish good, open working relationships with clients early in the project and maintain them throughout by communicating effectively.

Communication takes many forms, from written (memos, reports) to verbal (phone calls, face-to-face conversations) to visual (presentation slides, diagrams). The analyst must be able to master as many forms of communication as possible. Oral communication and listening skills are considered by many information system professionals as the most important communication skills analysts need to succeed. Interviewing skills are not far behind. All types of communication, however, have one thing in common: They improve with experience. The more you practice, the better you get. Some of the specific types of communication we will mention are interviewing and listening, the use of questionnaires, and written and oral presentations.

INTERVIEWING, LISTENING, AND QUESTIONNAIRES   Interviewing is one of the primary ways analysts gather information about an information systems project. Early in a project, you may spend a large amount of time interviewing users about their work and the information they use. There are many ways to effectively interview someone, and becoming a good interviewer takes practice. It is important to point out that asking questions is only one part of interviewing. Listening to the answers is just as important, if not more so. Careful listening helps you understand the problem you're investigating and, many times, the answers to your questions lead to additional questions that may be even more revealing and probing than the questions you prepared before your interview.

Although interviews are very effective ways of communicating with people and obtaining important information from them, interviews can also be very expensive and time-consuming. Because questionnaires provide no direct means by which to ask follow-up questions, they are generally less effective than interviews. It is possible, however, though time-consuming, to call respondents and ask them follow-up questions. Questionnaires are less expensive to conduct because the questioner does not have to invest the same amount of time and effort to collect the same information using a questionnaire as he or she does in conducting an interview. For example, using a written questionnaire that respondents complete themselves, you could gather the same information from 100 people in one hour that you could collect from only one person in a one-hour interview. In addition, questionnaires have the advantage of being less biased in how the results are interpreted because the questions and answers are standardized. Creating good questionnaires is a skill that comes only with practice and experience.

WRITTEN AND ORAL PRESENTATIONS   At many points during the systems development process, you must document the progress of the project and communicate that progress to others. This communication takes the following forms:

- Meeting agenda
- Meeting minutes
- Interview summaries
- Project schedules and descriptions
- Memoranda requesting information, an interview, participation in a project activity, or the status of a project
- Requests for proposal from contractors and vendors

and a host of other documents. This documentation is essential to provide a written, not just oral, history for the project, to convey information clearly, to provide details needed by those who will maintain the system after you are off the project team, and to obtain commitments and approvals at key project milestones.

The larger the organization and the more complicated the systems development project, the more writing you will have to do. You and your team members will have to complete and file a report at the end of each stage of the systems development life cycle. The first report will be the business case for getting approval to start the project. The last report may be an audit of the entire development process. And at each phase, the analysis team will have to document the system as it evolves. To be effective, you need to write both clearly and persuasively.

As there are often many different parties involved in the development of a system, there are many opportunities to inform people of the project's status. Periodic written status reports are one way to keep people informed, but there will also be unscheduled calls for ad hoc reports. Many projects will also involve scheduled and unscheduled oral presentations. Part of oral presentations involves preparing slides, overhead transparencies, or multimedia presentations, including system demonstrations. Another part involves being able to field and answer questions from the audience.

How can you improve your communication skills? We have four simple yet powerful suggestions:

1. Take every opportunity to practice. Speak to a civic organization about trends in computing. Such groups often look for local speakers to present talks on top-

ics of general interest. Conduct a training class on some topic on which you have special expertise. Some people have found participation in Toastmasters, an international organization with local sections, a very helpful way to improve oral communication skills.

2. Videotape your presentations and do a critical self-appraisal of your skills. You can view videotapes of other speakers and share your assessments with each other.

3. Make use of writing centers located at many colleges as a way to critique your writing.

4. Take classes on business and technical writing from colleges and professional organizations.

■ WORKING ALONE AND WITH A TEAM

As a systems analyst, you must often work alone on certain aspects of any systems development project. To this end, you must be able to organize and manage your own schedule, commitments, and deadlines. Many people in the organization will depend on your individual performance, yet you are almost always a member of a team and must work with the team toward achieving project goals. Working with a team entails a certain amount of give and take. You need to know when to trust the judgment of other team members as well as when to question it. For example, when team members are speaking or acting from their base of experience and expertise, you are more likely to trust their judgment than when they are talking about something beyond their knowledge. For this reason, the analyst leading the team must understand the strengths and weaknesses of the other team members. To work together effectively and to ensure the quality of the group product, the team must establish standards of cooperation and coordination that guide their work.

There are several dimensions to the cooperation and coordination that influence team work. Table 1.2 lists the twelve characteristics of a high performance team (McConnell, 1996). The first characteristic is a shared vision, which allows each team member to have a clear understanding of the project's objectives. A shared vision helps team members keep their priorities straight and not allow small items of little significance to become overwhelming and distracting. To provide motivation for team members, the vision also needs to present a challenge to team members. The second characteristic, team identity, emerges as team members work together closely and begin to share a common language and sense of humor. Team identity can lead to the synergy of effort only possible when groups work together well.

Shared vision and team identity are important but they alone may not be enough for a team to actually accomplish something. The third characteristic of high performance teams

TABLE 1.2
Characteristics of a High Performance Team (McConnell, 1996)

1. Shared, elevated vision or goal
2. Sense of team identity
3. Result-driven structure
4. Competent team members
5. Commitment to the team
6. Mutual trust
7. Interdependence among team members
8. Effective communication
9. Sense of autonomy
10. Sense of empowerment
11. Small team size
12. High level of enjoyment

is how the teams are organized. A result-driven structure is one that depends on clear roles, effective communication systems, means of monitoring individual performance, and decision making based on facts rather than emotions. Choosing the right people for the team is the fourth characteristic. McConnell (1996) reports that team performance may differ by as much as a factor of 5, depending only on the skills and attitudes of a team's members. Although the skills of each team member are important determinants of how well the team will perform, all members must be committed to the team, the fifth characteristic of high performance teams. A group of the best and brightest individuals, committed only to their own self-interests, cannot outperform a true team of lesser talents who are genuinely committed to each other and to their joint effort.

The next five characteristics of high performance teams all have to do with how the team members interact with each other. It is very important that team members develop genuine trust for each other. The need for trust is why you see so many team-building exercises; for example, an individual falls backwards into the arms of a fellow team member, not knowing if the other person is really there but trusting that he or she will be. Similarly, members of high-performance teams work interdependently, relying on each others' strengths; develop effective means of communication; give each team member the autonomy to do whatever he or she believes is best for the team and for the project; and empower each team member.

All of these high performance characteristics seem to work best, according to McConnell (1996), in small teams no larger than eight to ten people. Finally, it is important that teams have fun. Enjoying working together leads to increased team cohesiveness, which has been shown to be a key ingredient of team productivity (Lakhanpal, 1993).

▪ FACILITATING GROUPS

Sometimes you need to interact with a group in order to communicate and receive information. Let's talk about the Joint Application Design (JAD) process in which analysts actively work with groups during systems development. Analysts use JAD sessions to gather systems requirements and to conduct design reviews. The assembled group is the most important resource the analyst has access to during a JAD and you must get the most out of that resource; successful group facilitation is one way to do that. In a typical JAD, there is a trained session leader running the show. He or she has been specially trained to facilitate groups, to help them work together, and to help them achieve their common goals. Facilitation necessarily involves a certain amount of neutrality on the part of the facilitator. The facilitator must guide the group without being part of the group and must work to keep the effort on track by ferreting out disagreements and helping the group resolve differences. Obviously, group facilitation requires training. Many organizations that rely on group facilitation train their own facilitators. Figure 1-4 lists some guidelines for running an effective meeting, a task that is fundamental to facilitating groups.

▪ MANAGING EXPECTATIONS

Systems development is a change process, and any organizational change is greeted with anticipation and uncertainty by organization members. Organization members will have certain ideas, perhaps based on their hopes and wishes, about what a new information system will be able to do for them; these expectations about the new system can easily run out of control. Ginzberg (1981) found that successfully managing user expectations is related to successful systems implementation. For you to successfully manage expectations, you need to understand the technology and what it can do. You must understand the work flows that the technology will support and how the new system will affect them. More important than understanding, however, is your ability to communicate a realistic picture of the new system and what it will do for users and managers. Managing expectations begins with the development of the business case for the system and extends all the way through training people to use the finished system. You need to educate those who have few expectations as well as temper the optimism of those who expect the new system to perform miracles.

- Become comfortable with your role as facilitator by gaining confidence in your ability, being clear about your purpose, and finding a style that is right for you.
- At the beginning of the meeting, make sure the group understands what is expected of them and of you.
- Use physical movement to focus on yourself or on the group, depending on which is called for at the time.
- Reward group member participation with thanks and respect.
- Ask questions instead of making statements.
- Be willing to wait patiently for group members to answer the questions you ask them.
- Be a good listener.
- Keep the group focused.
- Encourage group members to feel ownership of the group's goals and of their attempts to reach those goals.

FIGURE 1-4   *Some Guidelines for Running Effective Meetings*
(Adapted from Option Technologies, Inc.[1992])

SYSTEMS ANALYSIS AS A PROFESSION

Even though systems analysis is a relatively new field, those in the field have established standards for education, training, certification, and practice. Such standards are required for any profession.

Whether or not systems analysis is a profession is open to debate. Some feel systems analysis is not a profession because it simply has not been around long enough to have established the rigorous standards that define a profession. Others feel that at least some standards are already in place. There are guidelines for college curricula and there are standard ways of analyzing, designing, and implementing systems. Professional societies that systems analysts may join include the Society for Information Management, the Association of Information Technology Professionals, and the Association for Computing Machinery (ACM). There is a Certified Data Processing Certificate (CDP) exam, much like the Certified Public Accountant (CPA) exam, that you can take to prove your competency in the field, although, unlike the CPA certificate, very few jobs and employers in the IS field require you to have the CDP certificate. Codes of ethics to govern behavior also exist. In this section, we will discuss several aspects of a systems analyst's job: standards of practice, the ACM code of ethics, and career paths for those choosing to become systems analysts.

■ STANDARDS OF PRACTICE

Standard methods or practices of performing systems development are emerging that make systems development less of an art and more of a science. Standards are developed through education and practice and spread as systems analysts move from one organization to another. We will focus here on four standards of practice: an endorsed development methodology, approved development platforms, well-defined roles for people in the development process, and a common language.

There are several different development methodologies now being used in organizations. Although there is no standardization of a single methodology across all organizations, a few prominent methodologies are in common use. An *endorsed development methodology* lays out specific procedures and techniques to be used during the development process. These standards are central to promoting consistency and reliability in methods across all of an organization's development projects. Some methodologies are spread through the work of well-known consultants, such as James Martin's Information Engineering™. Others are spread through major consulting firms, such as Andersen Consulting's Foundation/1™.

Closely associated with endorsed methodologies are approved development platforms. Some methodologies are closely tied to platforms, as is Andersen's methodology and their Foundation® CASE tool. Other methodologies are more adaptable and can work in close accordance with development platforms that exist in the organization, such as database management systems and 4GLs. The point is that organizations, and hence the analysts who work for them, are standardizing around specific platforms, and standards for development emerge from this standardization.

Roles for the various people involved in the development process are also becoming standardized. End-users, managers, and analysts are each assigned certain responsibilities for development projects. The training that analysts receive in college, on their first jobs, and during their interactions with other analysts, combine to create a gestalt of the analyst's job. For example, as you study this book and talk about systems development in your class, you are forming certain ideas about what systems analysts do and how systems are developed in organizations. Your ideas are also shaped and reinforced by the other IS courses you take in college. Once you get your first job, you will receive additional training and you will adjust your understanding of systems analysis accordingly. As you gain experience working on projects and interacting with other analysts, who may have been trained at other universities and in other organizations, your ideas will continue to change and grow, but the basic core of what systems analysis means to you will have been established. Many of the experiences you have on the job will reinforce much of what you have already learned about systems analysis. When you leave an organization and go to work elsewhere, you will carry your understanding of systems analysis with you. Over time, as you and other analysts change jobs and move from one organization to another, what it means to be an analyst becomes standardized across organizations, and the standards of practice in the field help define what it means to be an analyst.

Another factor moving the job of the systems analyst toward professionalism is the development of a common language analysts use to talk to each other. Analysts communicate on the job, at meetings of professional societies, and through publications. As analysts develop a special language for communication among themselves, their language becomes standardized. For example, since the late 1970s, systems analysts have begun to rely on data flow diagramming as a means of communication. There are now two primary standards for data flow diagramming: Yourdon, and Gane and Sarson. In time, only one may be used. Other examples of communication becoming standardized include the widespread use of common programming languages such as COBOL and C and the spread of SQL as the language of choice for data definition and manipulation for relational databases. As their common language develops, analysts become more cohesive as a group—a characteristic of professions.

■ ETHICS

The ACM is a large professional society made up of information system professionals and academics. It has over 85,000 members. Founded in 1947, the ACM is dedicated to promoting information processing as an academic discipline and to encouraging the responsible use of computers in a wide range of applications. Because of its size and membership, it has much influence in the information systems community. The ACM has developed a code of ethics for its members called the "ACM Code of Ethics and Professional Conduct." The full statement is reproduced in Figure 1-5. The code applies to all ACM members and directly applies to systems analysts.

Note the emphasis in the Code on personal responsibility, on honesty, and on respect for relevant laws. Notice also that compliance with a code of ethics such as this one is voluntary, although article 4.2 calls for, at a minimum, peer pressure for compliance. No one can force an information systems professional to follow these guidelines. However, it is voluntary compliance with the guidelines that makes someone a professional in the first place. Notice that for leaders there is the burden of educating non-IS professionals about computing—about what computing can and cannot do. The Code also expresses concern for the quality of work life and for protecting the dignity and privacy of others when performing professional work, such as developing information systems.

Association for Computing Machinery Professional Code of Ethics

Preamble

These statements of intended conduct are expected of every member (voting members, associate members, and student members) of the Association for Computing Machinery (ACM). Section 1.0 consists of fundamental ethical considerations; section 2.0 includes additional considerations of professional conduct; statements in 3.0 pertain to individuals who have a leadership role; and section 4.0 deals with compliance. ACM shall prepare and maintain an additional document for interpreting and following this Code.

(1.0) General Moral Imperatives

(As an ACM member I will . . .)
(1.1) Contribute to society and human well-being.
(1.2) Avoid harm to others.
(1.3) Be honest and trustworthy.
(1.4) Be fair and take action not to discriminate.
(1.5) Respect property rights (Honor copyrights and patents; give proper credit; not steal, damage, or copy without permission).
(1.6) Respect the privacy of others.
(1.7) Honor confidentiality

(2.0) Additional Professional Obligations

(As an ACM computing professional I will . . .)
(2.1) Strive to achieve the highest quality in the processes and products of my work.
(2.2) Acquire and maintain professional competence.
(2.3) Know and respect existing law pertaining to my professional work.
(2.4) Encourage review by peers and all affected parties.
(2.5) Give well-grounded evaluations of computer systems, their impacts, and possible risks.
(2.6) Honor contracts, agreements, and acknowledged responsibilities.
(2.7) Improve public understanding of computing and its consequences.

(3.0) Organizational Leadership Imperatives

(As an organizational leader I will . . .)
(3.1) Articulate social responsibilities of members of the organizational unit and encourage full participation in these responsibilities.
(3.2) Shape information systems to enhance the quality of working life.
(33) Articulate proper and authorized uses of organizational computer technology and enforce those policies.
(3.4) Ensure participation of users and other affected parties in system design, development and implementation.
(3.5) Support policies that protect the dignity of users and others affected by a computerized system.
(3.6) Support opportunities for learning the principles and limitations of computer systems.

(4.0) Compliance with Code

(4.1) I will uphold and promote the principles of this Code.
(4.2) If I observe an apparent violation of this Code, I will take appropriate action leading to a remedy.
(4.3) I understand that violation of this Code is inconsistent with continued membership in the ACM.

FIGURE 1-5 *ACM Code of Ethics and Professional Conduct, Revision Draft No. 19 (9/19/91). (© Association for Computing Machinery, reprinted with permission)*

Though not written specifically for systems analysts, the ACM Code of Ethics can easily be adapted to the systems analysis job. Many systems development projects deal directly with many of the issues addressed in the Code: privacy, quality of work life, user participation, and managing expectations. When an analyst must confront one or more of these issues, the Code can be used as a guide for professional conduct.

■ CAREER PATHS

A typical first job for a recent college graduate who wants to become a systems analyst is as an analyst or programmer/analyst trainee for a corporation or large consulting firm. Other typical entry-level opportunities are as

- End-user support specialist, assisting non-IS professional and clerical staff to better use computer resources
- Decision support analyst, in which you design database queries and data analysis routines to support business analysis and decision making, often for one department, such as market research or investments
- Trainer, in which you prepare and conduct various classes on information systems and technologies
- Computer technology sales and customer support, in which you either sell hardware, software, or services or support the sales staff by installing technology and responding to customer questions

Larger firms usually have their own intensive training programs to instruct trainees in the way the firm develops and maintains systems. Every firm handles systems development a little bit differently. Once trained, the entry-level systems analyst begins a career path, which differs from one firm to another.

For example, in one particular firm where the authors have done research—which we'll call the JKL Corporation—an analyst typically begins as an assistant analyst. The qualifications for this position are a bachelor's degree with little or no experience. Typically after two years of work, the assistant analyst is promoted to an associate analyst. Another two to three years of work experience qualify the associate analyst to become an analyst. The next two steps up the ladder are senior analyst and staff analyst. Of course, not everyone makes it to staff analyst. Many analysts leave after one or two promotions to work for other firms or settle into a position along the way.

At this point in his or her career at JKL Corp., the staff analyst must decide whether he or she wants to pursue a management ladder or a technical ladder. Analysts who are more technically oriented might prefer not to follow a management career path (see Figure 1-6). The management ladder continues upward with the positions of project leader, project manager, and manager. A manager is responsible for planning, developing, and implementing information technology at JKL Corp., for training and staff development, and for establishing and managing budgets. The technical ladder continues upward with the positions of consultant, staff consultant, and senior staff consultant. A senior staff consultant is responsible for major research and development projects. Creativity and vision are necessary for successful performance as a senior staff consultant.

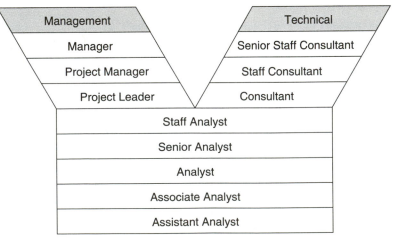

FIGURE 1-6   *Information Systems Career Ladder for the JKL Corp.*

It should be pointed out that not everyone hired by JKL Corp. starts out as an assistant analyst and works his or her way up in the corporation. Some employees are hired directly into higher-level jobs, based on their work experience and education. For example, someone with a master's degree and little or no experience can be hired directly into the associate analyst position. On the other hand, a person with a bachelor's degree and 14 or more years of proven experience can be hired directly into the manager's position.

Besides progression within the IS organization, some IS professionals choose to make a lateral move into general management positions and then return to senior IS management positions later in their careers. IS is an excellent entry point into any general business career since the process of systems analysis and design gives you a thorough understanding of the business, and your acquired general business knowledge is extremely valuable in general management positions.

For those who enter the consulting world, career progression is typically geared toward becoming a partner or principle of the business. These positions often focus on generating business and maintaining customer relationships. Another interesting option for people with more of an entrepreneurial spirit is to start your own business. You may develop expertise in a certain area or have a unique idea for some software tool that can flourish only in your own company. Many niche products, such as specialized data analysis tools or systems for particular types of functions or businesses, exist in the information systems field. Many of these were created by systems analysts who, after years of experience and continuing development, got a bright idea, networked with contacts, acquired funding, and started their own business. This is one way to become president of the company!

THE LIFE CYCLE OF A SYSTEM

In order to develop a computerized information system it is necessary for the process of development to pass through a number of distinct stages. The various stages in the systems life cycle are shown in Figure 1-7. These stages are completed in sequence. The project cannot progress from one stage to the next until it has completed all the required work of that stage. In order to ensure that a stage is satisfactorily completed some 'deliverable' is produced at the stage end. Generally this is a piece of documentary evidence on the work carried out. Successful completion of the stage is judged by the documentation. This is known as the 'exit criterion' for the stage.

It is common for some of the tasks carried out during a stage of the process to be initially unsatisfactory. This should come to light when the exit criteria are considered. The relevant tasks will need to he redone before exit from the stage can be made. Although 'looping' within a stage is commonplace, once a stage has been left it should not be necessary to return to it from a later stage. This structure—a linear development by stages, with deliverables and exit criteria—enables the project to be controlled and managed. The benefits of this staged approach are:

- Subdivision of a complex, lengthy project into discrete chunks of time, makes the project more manageable and thereby promotes better project control.

- Although different parts of a project may develop independently during a stage, the parts of the project are forced to reach the same point of development at the end of the stage. This promotes coordination between the various components of large projects.

- The deliverables, being documentation, provide a historical trace of the development of the project. At the end of each stage the output documentation provides an initial input into the subsequent stage.

- The document deliverables are designed to be communication tools between analysts, programmers, users and management. This promotes easy assessment of the nature of the work completed during the stage.

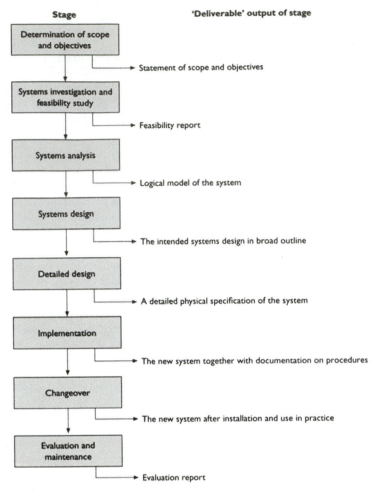

FIGURE 1-7   *Stages in the Life Cycle of a Systems Project*

- The stages are designed to be 'natural' division points in the development of the project.
- The stages allow a creeping commitment to expenditure during the project. There is no need to spend large sums of money until the previous stages have been satisfactorily completed (see Figure 1-8).

The approach progresses from the physical aspects of the existing system (systems investigation) through logical analysis (systems analysis) and logical design (systems design) on to the physical aspects of the new system (detailed design, implementation and evaluation).

■ STAGE 1 DETERMINATION OF SCOPE AND OBJECTIVES

Before an analyst can attempt to undertake a reasonable systems investigation, analysis and design there must be some indication given of the agreed overall scope of the project. The documentation provided on this acts as the analyst's initial terms of reference. This may be provided by the steering committee or written by the analyst and agreed by the committee. Either way it delimits the analyst's task.

The statement of scope and objectives will indicate an area to be investigated, such as sales order processing. It will also specify a problem or opportunity that the analyst should have in mind when investigating this area. For instance, sales order processing might be perceived to be too slow and the company fears that it is losing customers. The document should also specify a date by which the feasibility report (see Stage 2) is to be produced and the budgeted cost allowable for this.

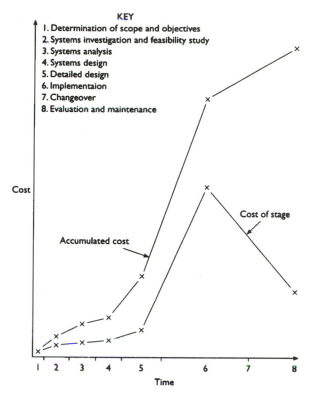

KEY
1. Determination of scope and objectives
2. Systems investigation and feasibility study
3. Systems analysis
4. Systems design
5. Detailed design
6. Implementaion
7. Changeover
8. Evaluation and maintenance

Cost

Cost of stage

Accumulated cost

Time

FIGURE 1-8   *Cost Profile of a Typical Systems Project*

■ STAGE 2 SYSTEMS INVESTIGATION AND FEASIBILITY STUDY

The output of this stage is a report on the feasibility of a technical solution to the problems or opportunities mentioned in the statement of scope and objectives in Stage 1. More than one solution may be suggested. The solution(s) will be presented in broad outline. An estimate of the costs, benefits and feasibility associated with each will be included. The purpose of the report is to provide evidence for the steering committee to decide on whether it is worth going ahead with any of the suggestions. If the whole project is dropped at this stage there will have been very little cost to date (sunk cost) (see Figure 1-8).

In order to establish the feasibility of a future technical system it will be necessary for the analyst to investigate the current system and its work practices. This will provide evidence for the functions that the new system will perform even in the case of substantial redesign. The analyst will need to interview users and view existing documentation. The information collected during this stage will be useful for the next stage of the life cycle as well.

■ STAGE 3 SYSTEMS ANALYSIS

Provided that the project has been given the 'go ahead' as a result of the feasibility study the next task for the analyst is to build a <u>logical model</u> of the existing system. This will be based partly on information collected during the stage of systems investigation and partly on new information gathered from the existing system.

The purpose of this stage is to establish what has to be done in order to carry out the functioning of the existing system. This 'what has to be done' is not to be confused with what actually happens in the existing physical system. That is, it is not to be confused with who does what to which document before transferring it from one department to another, or who provides what information for whom on whose authorization. Rather, the central question to be asked is 'what, logically speaking, must be done in order to satisfy the objectives and functions of the system under investigation?' This will involve a decomposition of the functions of the system into their logical constituents and the production of a logical model of the processes and data flows necessary to perform these. This is called process analysis. The logical model will be illustrated by data flow diagrams and structured techniques for specifying algorithms. Incorporated into this model will be any additional logical requirements to be made of the new system. No decisions on the way the system will be physically realized should be apparent from this model.

The processes will be fed by data. This data will relate to entities of interest to the organization. These entities will stand in relation to one another. For instance, data will be held on **EMPLOYEES** and the **DEPARTMENTS** in which they work. Thus **WORK** is a relationship between **EMPLOYEES** and **DEPARTMENTS**. These entities and relationships are combined in a data model of the organization. This procedure is called data analysis.

The output of this stage will be a logical process model as revealed by data flow diagrams, together with the specification of the process algorithms, data dictionary and a data model. This output is reviewed by management and users. Agreement needs to be established as to whether the model is a reflection of the logic of the problem area.

■ STAGE 4 SYSTEMS DESIGN

Once the analysis is complete the analyst has a good idea of what is logically required of the new system. There will be a number of ways that this logical model can be incorporated into a physical design. For instance, are the data stores to be implemented as a series of files or is there to be a database? If a database is the chosen route should this be centralized or distributed? The data flow diagrams will reveal the processes and the data flows between them. How many of the processes are to be incorporated into a computer system and how many are to remain manual? Of those to be computerized which are to be run under batch processes and which interactively online? Is the computerized system to be centralized or distributed?

There will not be one correct answer to these questions. Rather, there will be a range of alternative designs. Each of these will have different cost and efficiency implications. Some will yield more computerized facilities than others. Structured tools, such as data flow diagrams, enable these design alternatives to be clearly identified. They also allow the various options to be presented in a manner that requires little technical expertise in order to understand them.

The analysis will suggest two or three design alternatives to management together with their implications. Management will then decide amongst them. Often these alternatives will reflect a low-, a medium-, and a high-cost solution to the problem. The first will provide a system that is very basic. The second alternative will incorporate more facilities while the third may go beyond this to illustrate the full potential of extensive computerization.

By the end of this stage the attention of the analyst is turning away from purely logical considerations to the various ways, in general terms, that the logical model can be physically implemented. This stage ends with a choice between the alternatives presented to management.

■ STAGE 5 DETAILED DESIGN

At some point, merely looking at the logical requirements of a system or at broad-outline design solutions will no longer do. Detailed physical specifications need to be made so that the system can be purchased, built and installed. There are a number of distinct areas that must be considered.

1. Programs will need to be written so that the computer can perform the various functions required of it. These programs will be coded by programmers, who need a clear statement of the task to be programmed. Structured tools make possible clear specifications from which the program is written. They also enable the programs to be easily testable and amendable if necessary.

2. Hardware requirements must be specified that, together with the programs, will allow the computer system to perform its tasks efficiently. These requirements for terminals, disk drives, central processing units, cables and so on must be detailed enough to allow the purchasing department to obtain the items.

3. The structure of the database or system of files will also be specified.

4. A schedule for the implementation of the system will be derived at this stage. This will ensure that during implementation all the various activities are coordinated and come together at the right time.

These areas can be summarized as software, hardware, data storage and the schedule for implementation. Two other threads will run through consideration of these areas. The first is security. The system must be designed to ensure maximum reliability in secure, complete, accurate and continuous processing. The second is the user-machine interface. Unless this is designed with the characteristics of the tasks and the users in mind it is unlikely that the system will be fully effective in meeting its objectives.

The systems specification is a highly detailed set of documents covering every aspect of the system. From this it is possible to estimate costs accurately. The specification is finally ratified by senior management or the steering committee. Once it is agreed, large sums of the project budget can be spent on major purchases and programmers' time.

■ STAGE 6 IMPLEMENTATION

During implementation the system as specified is physically created. The hardware is purchased and installed. The programs are written and tested individually. As programs often interact they will also be tested together.

The database or file structure is created and historic data from the old system (manual or computer) is loaded. Staff are trained to use the new system. The procedures that will govern the operation of the new system are designed and documentation detailing these is then drafted. Particular attention will be paid to security features surrounding the conversion of existing files, whether manual or computer-based, to the new system.

The system is formally tested and accepted before changeover.

■ STAGE 7 CHANGEOVER

Changeover is that time during which the old system is replaced by the newly designed computer system. This period may be short if, at the time the new system starts running, the old system is immediately discarded.

Alternative methods of changeover exist. The old system can be run in parallel with the new. Although expensive in labour costs this method does have the advantage that if the new system fails there is a backup system to rely on. The old and the new systems can also be compared with one another for discrepancies in their performance and output. Another approach is to run a pilot scheme. This involves running a small version of the system before the full systems implementation is carried out. The way that the pilot system functions allows identification of any errors and shortcomings that will be encountered in the full system. These will be involved in the life cycle of a project. No matter how extensive the planning has been, no matter how rigorous the systems testing, there are always unexpected problems during the first few days or weeks of use of the new system. The problems should be minor if the preceding stages of the project have been carried out according to correct standards. These troubles may be technical or they may result from the use of the system for the first time by inexperienced (though trained) company personnel. After the system has 'settled down' the next phase of the life cycle is entered.

■ STAGE 8 EVALUATION AND MAINTENANCE

By now the system is running and in continual use. It should be delivering the benefits for which it was designed and installed. Any initial problems in running will have been rectified. Throughout the remainder of the useful life of the system it will have to be maintained if it is to provide a proper service.

The maintenance will involve hardware and software. It is customary to transfer the maintenance of the hardware to the manufacturer of the equipment or some specialist third-party organization. A maintenance contract stipulating conditions and charges for maintenance is usual. The software will need to be maintained as well. This will involve correcting errors in programs that become apparent after an extended period of use. Programs may also be altered to enable the machine to run with greater technical efficiency. But by far the greatest demand on programmers' time will be to amend and develop existing programs in the light of changes in the requirements of users. Structured techniques of design and programming allow these changes to be made easily.

It is customary to produce an evaluation report on the system after it has been functioning for some time. This will be drawn up after the system has settled into its normal daily functioning. The report will compare the actual system with the aims and objectives that it was designed to meet. Shortcomings are identified. If these are easily rectified then changes will be made during normal maintenance. More major changes may require more serious surgery. Substantial redesign of parts of the system may be necessary. Alternatively, the changes can be incorporated into a future system.

THE STRUCTURED APPROACH AND THE LIFE CYCLE

Structured systems analysis and design both define various stages that should be undertaken in the development of a systems project. In the life cycle the structured techniques and tools are used in analysis and design. Their benefits are realized throughout the project in terms of better project control and communication, and during the working life of the system in terms of its meeting user requirements and the ease with which it can be modified to take into account changes in these requirements.

The philosophy of the approach distinguishes it from more traditional methods used in analysis and design. Central to this is the idea that a logical model of the system needs to be derived in order to be able to redesign and integrate complex systems. This is evident in the stages of systems analysis and design. The detailed tools all follow from and through this central idea. Table 1.3 summarizes these as applied to the stages of the life cycle.

■ PROTOTYPING

Prototyping is an alternative approach to traditional methods of project management. These regard the development of a project as proceeding in a series of well-defined stages with deliverables at the end of each stage. Structured systems analysis and design is a typical example of this linear life-cycle approach. In contrast, prototyping views the development of computer systems as proceeding through a number of iterative loops. At each stage some version of the system is prepared for assessment and alteration at a later stage.

■ RAPID APPLICATIONS DEVELOPMENT

Rapid Applications Development (RAD) grew out of the recognition that businesses need to respond quickly to changing, often uncertain, environments in the development of their information systems. RAD is directly opposed to the traditional life-cycle approach with its emphasis on complete linear development of a system and concentration on technical perspectives. RAD borrows from other approaches and uses prototyping, participation, and CASE tools as well as other formal techniques. It recognizes the importance of gaining user, particularly senior management, involvement in its evolutionary approach to information systems development. RAD was first separately identified and introduced by James Martin (1991). His exposition set the methodology clearly within his Information Engineering approach to the development of business information systems. Now, though, the term Rapid Applications Development is used much more loosely to encompass any approach which emphasizes fast development of systems. Within the UK a consortium of systems developers are attempting to define a framework of standards for RAD called the Dynamic Systems Development Method (DSDM).

TABLE 1.3
Stages of the Life Cycle

Stages	Purpose	Comments
Determination of scope and objectives	To establish the nature of the problem, estimate its scope and plan feasibility study	
Systems investigation and feasibility study	To provide a report for management on the feasibility of a technical solution	Involves the analyst in investigation of the existing system and its documentation. Interviews used
Systems analysis	To provide a logical model of the data and processes of the system	Use of data flow diagrams, entity relationship models, structured English, logic flowcharts, data dictionaries
System design	To provide outline solutions to the problem	Automation boundaries indicated on data flow diagrams, suggestions offered on type of systems for example, centralized v distributed, file v database, Cost estimates provided
Detailed design	To provide a detailed specification of the system from which it can be built	Programs specified using hierarchical input process output (HIPO) and 03 hardware and file/database structures defined, cost estimates, systems test plan and implementation schedule designed
Implementation	To provide a system built and tested according to specification	Code programs, obtain and install hardware, design operating procedures and documentation, security/audit considerations test system, train staff, load existing data
Changeover	To provide a working system that has adequately replaced the old system	Direct, parallel, pilot or phased changeover
Evaluation and maintenance	To provide an evaluation of the extent to which the systems meets its objectives. Provide continuing support	Report provided. Ongoing adaptation of software/hardware to rectify errors and meet changing user requirements

Central to the concept of RAD is the role of clearly defined workshops. These should:

- involve business and information systems personnel;
- be of a defined length of time (typically between one and five days);
- be in 'clean rooms'—i.e. rooms set aside for the purpose, removed from everyday operations, provided with technical support, and without interruption;

- involve a facilitator who will be independent, control the meeting, set agendas, and be responsible for steering the meeting to deliverables;
- involve a scribe to record.

RAD has four phases:

1. *Requirements planning:* The role of Joint Requirements Planning (JRP) is to establish high level management and strategic objectives for the organization. The workshop will contain senior managers often cooperating in a cross-functional way. They will have the authority to take decisions over the strategic direction of the business. The assumption behind RAD is that JRP will drive the process from a high level business perspective rather than a technical one.

2. *Applications development:* Joint Applications Development (JAD) follows JRP and involves users in participation in the workshops. JAD follows a top-down approach and may use prototyping tools. Any techniques that can aid user design, especially data flow diagrams and entity modelling will be employed. I-CASE will be used at this stage. The important feature of applications development is that the JAD workshops short-circuit the traditional life-cycle approach which involves lengthy interviews with users and the collection of documentation often over considerable time periods. In JAD the momentum is not lost and several focused workshops may be called quite quickly.

3. *Systems construction:* The design specifications by JAD are used to develop detailed designs and generate code. In this phase graphical user interface building tools, database management system development tools, 4GLs, and back-end CASE tools are used. A series of prototypes are created which are then assessed by end-users which may result in further iterations and modifications. The various parts of the systems are developed by small teams, known as SWAT teams (Skilled With Advanced Tools). The central system can be built quickly using this approach. The focus of RAD is on the development of core functionality, rather than the 'whistles and bells' of the system—it is often claimed that 80% (the core) of the system can be built in 20% of the time.

4. *Cutover:* During cutover the users are trained and the system is tested comprehensively. The objective is to have the core functioning effectively. The remainder of the system can be built later. By concentration on the core and the need to develop systems rapidly within a 'timebox' the development process can concentrate on the most important aspects of the information system from a business perspective. If the process looks as though it is slipping behind schedule, out of its timebox, it is likely that the requirements will be reduced rather than the deadline extended.

Rapid Applications Development makes the assumption that:

- businesses face continual change and uncertainty;
- information requirements and therefore information systems must change to meet this challenge;
- information systems development should be driven by business requirements;
- it is important that information systems are developed quickly;
- prototyping and development tools are necessary to ensure quick responses;
- users should participate in development;
- the 'final system' does not exist.

RAD also assumes that consensus on aims and objectives can be reached and that problems are not 'messy.'

THE FEASIBILITY STUDY AND REPORT

One of the purposes for carrying out a systems investigation, perhaps *the* main purpose, is to establish the feasibility of introducing a computer system. Amongst other things this will provide some estimate of the likely costs and benefits of a proposed system. The reason for the study is to establish whether at as early a stage as possible the project is realistic. This must be determined with the minimum of expenditure. If the project turns out not to be feasible then all the time and money spent on the systems investigation will be 'down the drain.'

There is a conflict here. On the one hand, the earlier the feasibility study the less the money that will have been sunk, but the less likely it will be that the feasibility study gives an accurate estimate of costs and benefits. On the other hand, a more accurate assessment can only be made if more money is spent on the feasibility survey.

There is no completely satisfactory way of resolving this dilemma. In practice, the analyst is more likely to recommend an extensive feasibility study in more unusual and innovative projects. This is because the degree of uncertainty in success, costs and benefits is greater. However, many analysts become familiar with certain types of project such as the computerization of a standard accounting system. In these cases it will be possible to make reasonably accurate feasibility assessments quickly.

There is inevitably an element of guesswork at the feasibility stage (despite what some analysts might claim). The long history of notable failures of computerization is testimony to this fact as they can, in part, be put down to unrealistic feasibility studies. The more effort put into the study the less the guesswork. Sometimes parts of the stages of systems analysis and systems (high-level) design may be undertaken using the structured tools such as data flow diagrams and entity relationship models prior to producing a feasibility report. It is assumed here, in the case of Kismet, that the analyst has established enough information after investigation and initial interviews to have a thorough understanding of the present physical system and is able to recommend the feasibility of a computer system. The suggestion will be based on an understanding of the tasks to be performed, the volumes of transactions processed, and the types of information to be produced.

In looking at feasibility the analyst considers three main areas—economic, technical and organizational feasibility.

ECONOMIC FEASIBILITY

As with any project that an organization undertakes there will be economic costs and economic benefits. These have to be compared and a view taken as to whether the benefits justify the costs. If not, then the project is unlikely to be undertaken.

■ ECONOMIC COSTS

There are a number of different types of cost associated with a computer project. These are:

1. *Systems analysis and design:* The cost of the analyst must be taken into the calculation of the total cost of the project. Of course the analyst's costs in carrying out the stages up to and including the feasibility study will not be included in this calculation. These are already a sunk cost of the project.

2. *Purchase of hardware:* Alternatives to purchase such as leasing or renting may be considered here.

3. *Software costs:* These are often the hardest to estimate. Software may be written from scratch, developed using fourth-generation tools, or purchased, in the form of an applications package.

4. *Training costs:* Staff need to be trained to use the new system.

5. *Installation costs:* This may be a significant cost if new rooms have to be built, cables laid and work environments altered.

6. *Conversion and changeover costs:* These concern the loading of data from the existing system into the new system in a secure manner. There are also costs associated with the resulting changeover from the old to the new system.

7. *Redundancy costs:* If the purpose of computerization is to replace people with machines then redundancy money may have to be paid.

8. *Operating costs:*

   (a) maintenance costs for hardware and software;

   (b) costs of power, paper and so on;

   (c) costs associated with personnel to operate the new system—for example, computer center staff, data input clerks and so on.

■ ECONOMIC BENEFITS

These are often very varied. Some may be estimatable with a high degree of accuracy, others may be uncertain. Many benefits will be completely non-measurable. Examples of benefits are:

1. *Savings in labour costs:* These may be predictable allowing for uncertainties in future wage rates and so on.

2. *Benefits due to faster processing:* Examples of these might be a reduced debtor period as a result of speedier debtor processing, or reduced buffer stock due to better stock control. These may be estimatable.

3. *Better decision making:* Computerized information systems provide more targeted and accurate information quicker and cheaper than manual systems. This leads to better managerial decisions. It is generally not possible to put a figure on the value of better managerial decisions. Even if it were, it would be impossible to assign what percentage of this improvement was the result of better information and what was the result of other factors.

4. *Better customer service:* Once again, it will generally not be possible to estimate the economic benefits of either better customer service or more competitive services. This will be only one factor affecting customer choice.

5. *Error reduction:* The benefits of this may be estimatable if current losses associated with erroneous processing are known.

■ COMPARISON OF COSTS AND BENEFITS

Both costs and benefits occur in the future though not usually in the same future periods (see Figure 1-9). The costs occur largely during the initial stages of the systems development whereas the benefits occur later in the useful life of the system. These must be compared.

One method is to discount the future streams of costs and benefits back to the present by means of an assumed rate. This will be near to the prevailing rate of interest in the financial markets though its exact determination will depend on the project, the company undertaking the project, and the sector within which the company functions. This discount rate is arbitrary within certain limits.

All of the following factors:

• the non-measurable nature of some of the benefits;

• the fact that many of the benefits occur far into the uncertain future;

• the degree of arbitrariness associated with the choice of the cost/benefit comparison calculation;

mean that the estimation of economic feasibility must be taken with much reservation. It is tempting to regard the figure in the net present value calculation of the economic feasibility of the project as the 'hard' price of data on which a decision to continue the project can be made. This would be a mistake. It ignores not only the non-measurable nature of certain costs and benefits but also other aspects of feasibility covered in the following sections.

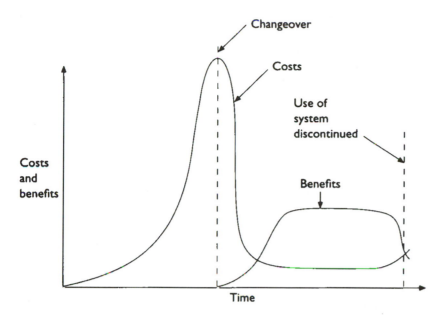

FIGURE 1-9   *The Time Profile of Costs and Benefits for a Typical Systems Life Cycle*

TECHNICAL FEASIBILITY

This is concerned with the technical possibility and desirability of a computer solution in the problem area. Some of the issues will overlap with points brought out in the discussion on costs and benefits in the previous section. Many suggestions are not technically impossible *per se*. Rather, it is a question of how much money an organization is prepared to invest in order to achieve a technical solution. The following categories are important in determining the technical feasibility of a project.

1. *Rule-governed tasks*: If the tasks undertaken in the area of investigation are not governed by rules but require discretion and judgment, it is unlikely that computerization will be feasible. For example, the processing of a sales order, the production of a summary of aged debtors report or the calculation of credit-worthiness on a points basis are all rule-governed. The selection of candidates for jobs is not. This could not be incorporated in a computer system (though new developments in expert systems raise the possibility that this might not always be so).

2. *Repetitive tasks*: If a task is performed only rarely then it may not be feasible to invest the time, effort and money to developing a program to carry it out. The tasks that are most suitable for computerization are those that are repetitive.

3. *Complex tasks*: If a complex task can be broken down into simple constituent tasks then it is generally easy and desirable to computerize.

4. *High degree of accuracy*: Humans are quite good at making quick estimates based on 'rule of thumb' assumptions in flexible circumstances. Computers are not. However, if a high degree of numerical accuracy is required then computers outstrip humans by both speed and low error rates.

5. *Speed of response*: Computer systems give fast responses if required and are designed to do so.

6. *Data used for many tasks*: Once data is inside a computer it is generally easy and cheap to use it repeatedly for different tasks.

ORGANIZATIONAL FEASIBILITY

Organizational feasibility or, as it is sometimes called, 'operational feasibility' concerns the viability of the proposed system within the operational and organizational environment. The issues to consider vary from organization to organization but the analyst is wise to address at least the following questions:

1. Does the organization for which the information system is to be supplied have a history of acceptance of information technology or has past introduction led to conflict? Some sectors are notorious for their opposition to computerization. For instance, in the UK the print industry unions fought an extended battle opposing the introduction of computer technology. Other sectors, such as banking, have a history of acceptance of and adaptation to information technology. A previous history of opposition to the introduction of computer systems may have taken the form of a formalized union opposition or it may have been revealed in the attitude of users. High levels of absenteeism and high turnover rates subsequent to a previous introduction of new technology are good indicators of future poor acceptance.

2. Will the personnel within the organization be able to cope with operating the new technology? It is unrealistic, for instance, to expect elderly staff, familiar with working within the existing manual system, to adapt readily to new technology no matter how much training is given.

3. Is the organizational structure compatible with the proposed information system? For example, a highly centralized autocratic management structure is generally not compatible with a distributed computer system. Decentralized systems inevitably lead to local autonomy and local management of computer resources. Similarly, if departments or divisions within an organization have a history of competing with one another rather than cooperating, it is unlikely that it will be easy to develop a successful integrated system.

These are all issues in the area of organizational behaviour and 'people problems.' Analysts often have a training in programming or other technical areas and it is easy for them to ignore this vital component of feasibility.

FEASIBILITY REPORT

A feasibility report will be written by the analyst and considered by management prior to allowing the project to continue further. It will go to the steering committee in the case of a large organization. In a smaller organization without a steering committee structure the report will be assessed by senior managers as part of their normal activities.

As well as providing information on the feasibility of the project the systems investigation will have provided much information that will also be incorporated in the feasibility report. In particular:

- The principal work areas for the project will have been identified.
- Any needs for specialist staff to be involved in the later stages of the project will have been noted.
- Possible improvement or potential for savings may have become apparent during the investigation.

Outline headings for a typical feasibility report are given in Figure 1-10.

Title page:	Name of project, report name, version number, author, date.
Terms of reference:	These will be taken from the statement of scope and objectives.
Summary:	This gives a clear concise statement of the feasibility study and its recommendations.
Background:	Statement of the reasons for initiation of the project, the background of the current system, how it features within the organization, how it figures in the organization's development plans, what problems it encounters.
Method of study:	Detailed description of the systems investigation including personnel interviewed and documents searched, together with any other channels of information. Assumptions made and limitations imposed.
Present system:	Statement of the main features of the current system including its major tasks, its staffing, its storage, its equipment, its control procedures, and the way it relates to other systems within the organization.
Proposed system(s):	Each proposed system, if there is more than one, is outlined. This will include a statement of the facilities provided. (Data flow diagrams and other charting techniques may be used as a pictorial representation of the proposal.) For each proposal its economic, technical and organizational feasibility will be assessed. Major control features will be included.
Recommendation:	The recommended system will be clearly indicated with reasons why it is preferred.
Development plan:	A development plan for the recommended system is given in some detail; this will include projected costs for future stages in the life cycle with estimates of the time schedule for each.
Appendix:	This will provide supporting material to the main report. It will include references to documents, summaries of interviews, charts and graphs showing details of transaction processing, estimates of hardware costs and so on. In fact, anything that is likely to be of use to those reading the report that will enable them to make a more informed decision will be included.

FIGURE 1-10   *The Contents of a Typical Feasibility Report*

Once the feasibility report has been accepted the project can proceed to the next stage. This is to provide an analysis from which a new system can be designed and implemented. Various tools and techniques are normally used in analysis, although there is nothing to stop the analyst using them in the stages of systems investigation. The various charts and diagrams can then be included in the feasibility report. This is tantamount to carrying out broad aspects of *systems* analysis and *systems* design (as opposed to *detailed* design) prior to the provision of the feasibility report. This makes possible a more comprehensive development of a proposal or range of proposals. It also allows better communication of these proposals within the feasibility report as the techniques used are designed for facilitating communication.

ANALYSIS

SYSTEMS INVESTIGATION

The analyst now must become thoroughly familiar with the existing system. In particular the analyst has to determine:

- the objectives of the existing system;
- how the existing system works;
- any legal, government, or other regulations that might affect the operation of the system;
- the economic and organizational environment within which the system lies and in particular any changes that are likely to occur.

Why should the analyst pay much attention to the workings of the existing system because, after all, is this system not deficient? Or else why would there be a need to replace it? There are a number of observations to make. First, although it is assumed that the problem is one that is amenable to a computerized solution this has not, as yet, been established. It may turn out that a change in existing manual procedures or organizational structure is the best way of solving the problem. This will only come to light after investigation of the existing system. Of course analysts may be blind to such alternatives. Analysts are trained to look for technical solutions. They may also have a vested commercial interest in a computerized solution so that it is easy for them to miss alternative solutions. There is, though, a second reason for extensively studying the existing system. This will give the analyst a thorough understanding of the nature of the activities to be incorporated in the final computerized system. No matter how weak the existing system is, it must function at some level of effectiveness. This will provide a rich source of information from which the analyst can work.

THE ANALYST'S CHANNELS OF INFORMATION

The analyst needs to obtain information about the existing system and its environment. There are five main sources that the analyst can use.

1. interviews
2. documentation
3. observation
4. questionnaires
5. measuring.

Statement of scope and objectives

Project name: Sales order processing—Kismet Ltd date: mm/dd/yy

Current problems:

The following problems have been identified:

1. The sales catalogue of prices and products used to price customer orders is often out of date. In particular, new items are not catalogued immediately and items that are now no longer stocked still appear. The problem is located in the time-consuming nature of the manual preparation of the catalogue from the inventory records.

2. Customer enquiries are sometimes difficult to deal with as records of orders are not stored in a form that is easily accessible.

3. Orders that cannot be immediately satisfied from current stock—that is, back orders—are not processed consistently.

4. Owing to the large number of documents flowing through the system the time taken to process an order may be days, even if all the goods are held in the warehouse.

5. Data within the system is generally stored in a way that makes it difficult for management to retrieve useful information. For instance, regular reports are time-consuming to produce and are often late, rendering them ineffective for control purposes or to aid medium-term strategies.

Objectives:
  To investigate initially the feasibility for computerization of the sales order processing, invoicing and stock systems.

Constraints:
  The entire project is to be budgeted for completion within six months at a cost of approximately $200,000.

Plan of action:
  Investigate fully the existing sales order processing, stock and invoicing systems. Investigate the feasibility of a computerized system as a solution to the current problems.

  Outline in general terms the recommended system(s) with costs.

  Produce a report on this feasibility within two weeks with a budget of $3,000.

FIGURE 2-1   *Statement of Scope and Objectives*

■ INTERVIEWS

This is the most important way in which an analyst will obtain information. Setting up interviews with key personnel at all levels in the organization ensures a rich and complete view of what is happening. Interviewing is more of an art than a mechanical technique. It improves with experience. There are, however, several guidelines that are recognized as being essential to successful interviewing.

First of all the analyst must have a clear purpose for each interview undertaken. This should be specified by the analyst as part of the preparation for the interview. It is not enough to define the purpose as 'attempting to find out more about such-and-such an area.' This will lead to a rambling interview. Rather, the analyst should establish the missing information that the interview is meant to supply. The analyst should prepare thoroughly for the interview by becoming familiar with technical terms that are likely to be used by the interviewees and with their positions and general responsibilities. The analyst should also outline a list of questions to be asked during the interview.

During the interview the analyst should:

- Explain at the beginning the purpose of the interview. This gives the interviewee a framework of reference for answering questions.

- Attempt to put the interviewee at ease.

- Go through the questions that were prepared. General questions should be asked first followed by more specific questions on each topic area. The analyst should always listen carefully to replies and be able to follow up answers with questions that were not in the original list. The analyst must always bear in mind the purpose of the interview and discourage time-wasting digressions.

- Never criticize the interviewee. The analyst is merely seeking information.

- Not enter into a discussion of the various merits or weaknesses of other personnel in the organization.

- Summarize points made by the interviewee at suitable stages in the interview.

- Explain the purpose of note taking or a tape recorder if used.

- Keep the interview short; generally 20 minutes or half an hour is sufficient.

- Summarize the main points of the interview at the end.

- Book a following interview, if required, with the interviewee at the end of the interview.

No checklist of guidelines is adequate to become a good interviewer. The list given should enable any serious pitfalls to be avoided.

PROBLEMS WITH THE INTERVIEW AS A CHANNEL OF INFORMATION The interview, though the most valuable tool for information gathering for the analyst, is limited in that:

1. The interviewee may refuse to cooperate with the interviewer through fear of job de-skilling, redundancy, or the inability to cope with the new technology as a result of computerization. This may take the form of a direct refusal to take part (unlikely), being vague in replies, or, by omission, continuing to let the analyst believe what the interviewee knows to be false.

2. The interviewee may feel that they should tell the analyst how the tasks that they carry out *should* be performed rather than how they actually *are* performed. It is common for people to cut corners, not follow works procedures, adopt alternative practices. All of these may be more efficient than the officially recommended practice but it is difficult for the interviewee to be honest in this area.

3. Clerical workers do tasks. They generally do not have to describe them and may not be articulate in doing so.

4. The analyst cannot avoid filtering all that the interviewee says through the analyst's model of the world. The analyst's background and preconceptions may interfere with the process of communication. One of the distinguishing marks of good interviewers is the ability to think themselves quickly into the interviewee's frame of mind. This almost therapeutic skill is not one that is usually developed through a training in computing.

■ DOCUMENTATION

Most business organizations, particularly large ones, have documentation that is of help to the analyst in understanding the way they work:

- Instruction manuals and procedures manuals provide a statement of the way that tasks are to be performed.

- Document blanks that are filled in by personnel within the organization and then passed between departments or stored for reference give the analyst an indication of the formal data flows and data stores.

- Job descriptions define the responsibilities of personnel.

- Statements of company policy provide information on overall objectives and likely changes.

- Publicity and information booklets for external bodies provide a useful overview of the way that a company works.

The problem with using documentation is that there is often a great deal of it, particularly in large organizations. The analyst has to read extensively in order to gather a small amount of useful information. Unlike interviews, where the analyst can direct the information that is provided by targeted questions, documents cannot be so easily probed. Finally, documentation may be out of date and the analyst has little way of knowing this. The last thing to be changed when a clerical procedure is altered is usually the documentation governing it. Despite these weaknesses documentation is a useful channel for information gathering.

■ OBSERVATION

Observation of employees performing activities within the area of investigation is another source of information for the analyst. Observation has the edge over the other methods of information gathering in that it is direct. The analyst wishes to understand the way that the existing system functions. Interviews provide reports from people of what they do, subject to all the distorting influences stated. Documents are an indication of what employees should be doing, which is not necessarily what they are doing. Only by observation does the analyst see directly how activities are performed.

However there are some notable drawbacks:

- It is extremely time-consuming for the analyst.

- When observed people tend to behave differently as compared to their behaviour unobserved—the 'Hawthorn effect'—thus devaluing the information obtained.

- Observation, unlike interviewing, does not reveal the beliefs and attitudes of the people involved.

Observation is, however, an important source for the analyst on informal information flows between individuals. These are often essential for the efficient execution of activities. They may not be obvious from interviews and would not appear in documentation.

■ QUESTIONNAIRES

Questionnaires are of only limited use in obtaining information for the purposes of investigating an existing system (as opposed to market research where they are essential). This is because:

- It is difficult to avoid misunderstandings on the part of respondents as they cannot gain clarification of a question on the questionnaire if it is judged to be vague or confusing.

- Questionnaires that are simple provide little information; questionnaires that are more ambitious are likely to be misunderstood.

- Response rates to questionnaires are often low.

- To create a good questionnaire the analyst often has to have more information about the system under investigation than the questionnaire could hope to provide in the first place.

Certain limited situations may make a questionnaire suitable. These usually occur when the number of people involved makes interviewing prohibitively expensive, the questions are generally simple, a low response rate is satisfactory, and the questionnaire is used to confirm evidence collected elsewhere.

In designing questionnaires it is important to:

- Keep questions simple, unambiguous and unbiased.
- Use multiple-choice questions rather than ask for comments. This makes the questionnaire both easier to answer and easier to analyze.
- Have a clear idea of the information that is required from the questionnaire.
- Make sure that the questions are aimed at the level of intellect and particular interests of the respondents.
- Avoid branching: for example, 'if your answer to question 8 was "yes" then go to question 23 otherwise go to question 19.'
- Make clear the deadline date by which the questionnaire is to be returned and enclose an addressed and prepaid envelope.

▪ MEASURING

Sometimes it is important to have statistical information about the workings of the existing system. The total number of sales ledger accounts and the activity of each will be of interest to the analyst who is looking at the possible computerization of an accounting system. The statistical spread as well as the gross figures may be relevant. For instance, with a sales order processing system not only may the average number of sales orders processed a day be of use to the analyst, but the pattern of these orders throughout the day and throughout the week may be of significance. Are there peaks and troughs or is it a constant flow?

APPROACHING THE INVESTIGATION

Although the foregoing channels provide the analyst with information it is necessary to have some plan or some framework within which to study the existing system.

▪ FLOW BLOCK DIAGRAMS

A flow block diagram may be developed at an early stage in the investigation to represent the system. Flow block diagrams show the important subsystems in an organization and the flows between them. They provide a good overview of a system within which more detailed investigation can occur. It is common for flow block diagrams to be based around the traditional functions of a business—sales, purchasing, manufacturing, stores, accounting, planning, control, and so on. A flow block diagram of Kismet is given in Figure 2-2.

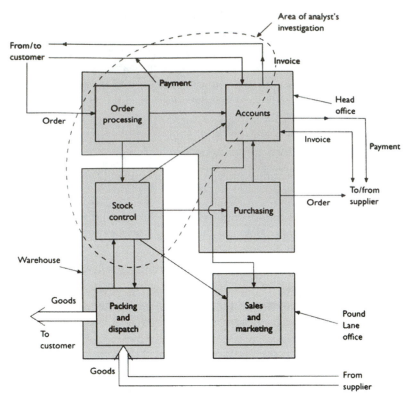

FIGURE 2-2   *A Flow Block Diagram of Kismet*

■ ORGANIZATION CHARTS

Organization charts show the various roles and their relationships within an organization. They are usually hierarchical in nature, reflecting relationships of control, decision flow and levels of managerial activity between the various elements of the hierarchy. The chart enables the analyst to establish key personnel for interview. An organization chart for Kismet is given in Figure 2-3.

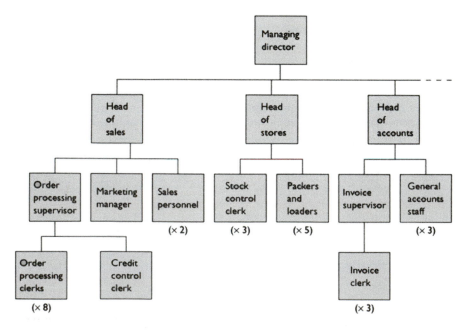

FIGURE 2-3   *An Organization Chart for Kismet*

■ TASK IDENTIFICATION

Within each subsystem the analyst will identify key tasks. A useful model to adopt is the system model (see Figure 2-4), where the task is regarded as a process for converting an input into an output. There may be intermediate storage requirements and there will be some control over the operation of the task. This gives the analyst a template by which a task can be investigated. Key questions that should be satisfied are:

- What different types of input are there into the task?
- And for each input:
  - What is the structure of the input?
  - Where does it come from?
  - What is the rate of input (how many per hour)?
  - Is it regular or are there peaks and troughs?
- What different types of output are there to the task?
- And for each output:
  - What is the structure of the output?
  - Where does it go to?
  - What is the rate of output (how many per hour) required?
  - Is it regular or are there peaks and troughs?
  - What is the purpose of the output?
- What is the logic of the process?
- Does it require discretion or is it rule-governed?
- What is the purpose of the process?
- What experience or training is necessary to become competent at it?
- What level of accuracy is required?
- What stores, files or records are consulted in performing the task?
- How often are these consulted?
- What indexes or keys are used for selecting the correct record?

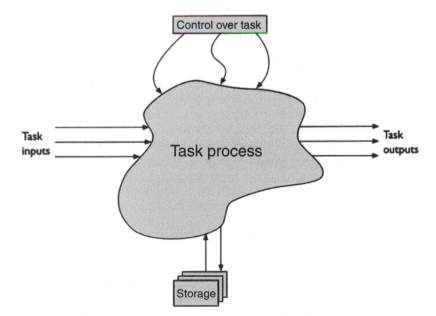

FIGURE 2-4   *A Systems Model of a Task (Task Template)*

- How many records are in the available record store?
- What types of control are exerted over the task?
- Who is responsible for each control?

After investigation the analyst should have a good understanding of:

- Who does what.
- Where it is done.
- Why it is done.
- When it is done.
- How it is done.

DETERMINING SYSTEM REQUIREMENTS

Systems analysis is the part of the systems development life cycle in which you determine how the current information system functions and assess what users would like to see in a new system. There are three subphases in analysis: requirements determination, requirements structuring, and alternative generation and choice.

In this section, you will learn about the beginning subphase of analysis—determining system requirements. Techniques used in requirements determination have evolved over time to become more structured and, as we will see in this section, current methods increasingly rely on the computer for support. We will first study the more traditional requirements determination methods including interviewing, using questionnaires, observing users in their work environment, and collecting procedures and other written documents. We will then discuss modern methods for collecting system requirements. The first of these methods is Joint Application Design (JAD). Next, you will read about how analysts rely more and more on information systems to help them perform analysis. As you will see, group support systems have been used to support systems analysis, especially as part of the JAD process. CASE tools are also very useful in requirements determination. Finally, you will learn how prototyping can be used as a key tool for some requirements determination efforts.

PERFORMING REQUIREMENTS DETERMINATION

As shown in Figure 2-5, there are three subphases to systems analysis: requirements determination, requirements structuring, and generating alternative system design strategies and selecting the best one. We will address these as three separate steps, but you should consider these steps as somewhat parallel and iterative. For example, as you determine some aspects of the current and desired system(s), you begin to structure these requirements or to build prototypes to show users how a system might behave. Inconsistencies and deficiencies discovered through structuring and prototyping lead you to explore further the operation of current system(s) and the future needs of the organization. Eventually your ideas and discoveries converge on a thorough and accurate depiction of current operations and what the requirements are for the new system. As you think about beginning the analysis phase, you probably wonder what exactly is involved in requirements determination.

■ THE PROCESS OF DETERMINING REQUIREMENTS

Once management has granted permission to pursue development of a new system (this was done at the end of the project identification and selection phase of the SDLC) and a project is initiated and planned, you begin determining what the new system should do. During requirements determination, you and other analysts gather information on what the system should do from as many sources as possible: from users of the current system, from observing users,

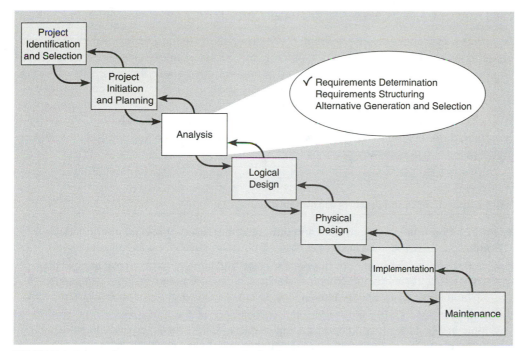

FIGURE 2-5   *Systems Development Life Cycle with Analysis Phase Highlighted*

and from reports, forms, and procedures. All of the system requirements are carefully documented and made ready for structuring.

In many ways, gathering system requirements is like conducting an investigation. Have you read any of the Sherlock Holmes or similar mystery stories? Do you enjoy solving puzzles? From these experiences, we can detect some similar characteristics for a good systems analyst during the requirements determination subphase. These characteristics include

- *Impertinence:* You should question everything. You need to ask such questions as: Are all transactions processed the same way? Could anyone be charged something other than the standard price? Might we someday want to allow and encourage employees to work for more than one department?

- *Impartiality:* Your role is to find the best solution to a business problem or opportunity. It is not, for example, to find a way to justify the purchase of new hardware or to insist on incorporating what users think they want into the new system requirements. You must consider issues raised by all parties and try to find the best organizational solution.

- *Relax constraints:* Assume anything is possible and eliminate the infeasible. For example, do not accept this statement: "We've always done it that way, so we have to continue the practice." Traditions are different from rules and policies. Traditions probably started for a good reason but, as the organization and its environment change, traditions may turn into habits rather than sensible procedures.

- *Attention to details:* Every fact must fit with every other fact. One element out of place means that the ultimate system will fail at some time. For example, an imprecise definition of who a customer is may mean that you purge customer data when a customer has no active orders; yet these past customers may be vital contacts for future sales.

- *Reframing:* Analysis is, in part, a creative process. You must challenge yourself to look at the organization in new ways. You must consider how each user views his or her requirements. You must be careful not to jump to this conclusion: "I worked on a system like that once—this new system must work the same way as the one I built before."

■ DELIVERABLES AND OUTCOMES

The primary deliverables from requirements determination are the various forms of information gathered during the determination process: transcripts of interviews; notes from observation and analysis of documents; analyzed responses from questionnaires; sets of forms, reports, job descriptions, and other documents; and computer-generated output such as system prototypes. In short, anything that the analysis team collects as part of determining system requirements is included in the deliverables resulting from this subphase of the systems development life cycle. Table 2.1 lists examples of some specific information that might be gathered during requirements determination.

TABLE 2.1
Deliverables for Requirements Determination

1. **Information collected from conversations with or observations of users:** interview transcripts, questionnaire responses, notes from observation, meeting minutes

2. **Existing written information:** business mission and strategy statements, sample business forms and reports and computer displays, procedure manuals, job descriptions, training manuals, flow charts and documentation of existing systems, consultant reports

3. **Computer-based information:** results from Joint Application Design sessions, transcripts or files from group support system sessions, CASE repository contents and reports of existing systems, and displays and reports from system prototypes.

These deliverables contain the information you need for systems analysis within the scope of the system you are developing. In addition, you need to understand the following components of an organization:

- The business objectives that drive what and how work is done
- The information people need to do their jobs
- The data (definition, volume, size, etc.) handled within the organization to support the jobs
- When, how, and by whom or what the data are moved, transformed, and stored
- The sequence and other dependencies among different data-handling activities
- The rules governing how data are handled and processed
- Policies and guidelines that describe the nature of the business and the market and environment in which it operates
- Key events affecting data values and when these events occur

As should be obvious, such a large amount of information must be organized in order to be useful. This is the purpose of the next subphase—requirements structuring.

From just this subphase of analysis, you probably already realize that the amount of information to be gathered could be huge, especially if the scope of the system under development is broad. The time required to collect and structure a great deal of information can be extensive and, because it involves so much human effort, quite expensive. Too much analysis is not productive and the term "analysis paralysis" has been coined to describe a systems development project that has bogged down in an abundance of analysis work. Because of the dangers of excessive analysis, today's systems analysts focus more on the system to be developed than on the current system. JAD and prototyping were developed to keep the analysis effort at a minimum yet still effective. Other processes have been developed to limit the analysis commitment even more, providing an alternative to the SDLC. One of these processes is called *Rapid Application Development (RAD)*. RAD relies on JAD, prototyping, and integrated CASE tools to be effective. Even RAD, as well as structured analysis methods, rely on a basic understanding of the business area served by an information system. Thus, before you can fully appreciate RAD, you need to learn about traditional fact-gathering techniques.

TRADITIONAL METHODS FOR DETERMINING REQUIREMENTS

At the core of systems analysis is the collection of information. At the outset, you must collect information about the information systems that are currently being used and how users would like to improve the current systems and organizational operations with new or replacement information systems. One of the best ways to get this information is to talk to the people who are directly or indirectly involved in the different parts of the organizations affected by the possible system changes: users, managers, funders, etc. Another way to find out about the current system is to gather copies of documentation relevant to current systems and business processes. In this section, you will learn about various ways to get information directly from stakeholders: interviews, questionnaires, group interviews, and direct observation. You will learn about collecting documentation on the current system and organizational operation in the form of written procedures, forms, reports, and other hard copy. These traditional methods of collecting system requirements are listed in Table 2.2.

TABLE 2.2
Traditional Methods of Collecting System Requirements

- Individually *interview* people informed about the operation and issues of the current system and needs for systems in future organizational activities
- Survey people via *questionnaires* to discover issues and requirements
- *Interview groups of people* with diverse needs to find synergies and contrasts among system requirements
- *Observe workers* at selected times to see how data are handled and what information people need to do their jobs
- *Study business documents* to discover reported issues, policies, rules, and directions as well as concrete examples of the use of data and information in the organization

■ INTERVIEWING AND LISTENING

Interviewing is one of the primary ways analysts gather information about an information systems project. Early in a project, an analyst may spend a large amount of time interviewing people about their work, the information they use to do it, and the types of information processing that might supplement their work. Other stakeholders are interviewed to understand organizational direction, policies, expectations managers have on the units they supervise, and other nonroutine aspects of organizational operations. During interviewing you will gather facts, opinions, and speculation and observe body language, emotions, and other signs of what people want and how they assess current systems.

There are many ways to effectively interview someone and no one method is necessarily better than another. Some guidelines you should keep in mind when you interview are summarized in Table 2.3.

First, you should prepare thoroughly before the interview. Set up an appointment at a time and for a duration convenient for the interviewee. The general nature of the interview should be explained to the interviewee in advance. You may ask the interviewee to think

TABLE 2.3
Guidelines for Effective Interviewing

Plan the Interview

- Prepare interviewee: appointment, priming questions

- Prepare checklist, agenda, and questions

Listen carefully and take notes (tape record if permitted)

Review notes within 48 hours of interview

Be neutral

Seek diverse views

about specific questions or issues or to review certain documentation to prepare for the interview. You should spend some time thinking about what you need to find out and write down your questions. Do not assume that you can anticipate all possible questions. You want the interview to be natural and, to some degree, you want to spontaneously direct the interview as you discover what expertise the interviewee brings to the session.

You should prepare an interview guide or checklist so that you know in which sequence you intend to ask your questions and how much time you want to spend in each area of the interview. The checklist might include some probing questions to ask as follow-up if you receive certain anticipated responses. You can, to some degree, integrate your interview guide with the notes you take during the interview, as depicted in a sample guide in Figure 2-6. This same guide can serve as an outline for a summary of what you discover during an interview.

The first page of the sample interview guide contains a general outline of the interview. Besides basic information on who is being interviewed and when, you list major objectives for the interview. These objectives typically cover the most important data you need to collect, a list of issues on which you need to seek agreement (for example, content for certain system reports), and which areas you need to explore, not necessarily with specific questions. You also include reminder notes to yourself on key information about the interviewee (for example, job history, known positions taken on issues, and role with current system). This information helps you to be personal, shows that you consider the interviewee important, and may assist you in interpreting some answers. Also included is an agenda for the interview with approximate time limits for different sections of the interview. You may not follow the time limits precisely but the schedule helps you cover all areas during the time the interviewee is available. Space is also allotted for general observations that do not fit under specific questions and for notes taken during the interview about topics skipped or issues raised that could not be resolved.

On subsequent pages you list specific questions; the sample form in Figure 2-6 includes space for taking notes on these questions. Because unanticipated information arises, you will not strictly follow the guide in sequence. You can, however, check off the questions you have asked and write reminders to yourself to return to or skip certain questions as the dynamics of the interview unfold.

CHOOSING INTERVIEW QUESTIONS   You need to decide what mix and sequence of open-ended and closed-ended questions you will use. Open-ended questions are usually used to probe for information for which you cannot anticipate all possible responses or for which you do not know the precise question to ask. The person being interviewed is encouraged to talk about whatever interests him or her within the general bounds of the question. An example is, "What would you say is the best thing about the information system you currently use to do your job?" or "List the three most frequently used menu options." You must react quickly to answers and determine whether or not any follow-up questions are needed for clarification or elaboration. Sometimes body language will suggest that a user has given an incomplete answer or is reluctant to divulge some information; a follow-up question might yield additional insight. One advantage of open-ended questions in an interview is that previously unknown information can surface. You can then continue exploring along unexpect-

Interview Outline	
Interviewee: *Name of person being interviewed*	Interviewer: *Name of person leading interview*
Location/Medium: *Office, conference room, or phone number*	Appointment Date: Start Time: End Time:
Objectives: *What data to collect* *On what to gain agreement* *What areas to explore*	Reminders: *Background/experience of interviewee* *Known opinions of interviewee*

Agenda:
Introduction
Background on Project
Overview of Interview
  Topics to Be Covered
  Permission to Tape Record
Topic 1 Questions
Topic 2 Questions
...
Summary of Major Points
Questions from Interviewee
Closing
	Approximate Time:
	1 minute
	2 minutes
	
	1 minute
	
	5 minutes
	7 minutes
	...
	2 minutes
	5 minutes
	1 minute

General Observations: *Interviewee seemed busy — probably need to call in a few days for follow-up questions since he gave only short answers. PC was turned off —probably not a regular PC user.*	
Unresolved Issues, Topics not Covered: *He needs to look up sales figures from 1996. He raised the issue of how to handle returned goods, but we did not have time to discuss.*	

(continues)

FIGURE 2-6   *Typical Interview Guide*

ed lines of inquiry to reveal even more new information. Open-ended questions also often put the interviewees at ease since they are able to respond in their own words using their own structure; open-ended questions give interviewees more of a sense of involvement and control in the interview. A major disadvantage of open-ended questions is the length of time it can take for the questions to be answered. In addition, open-ended questions can be difficult to summarize.

Closed-ended questions provide a range of answers from which the interviewee may choose. Here is an example:

> *Which of the following would you say is the one best thing about the information system you currently use to do your job (pick only one):*
> *a. Having easy access to all of the data you need*
> *b. The system's response time*
> *c. The ability to run the system concurrently with other applications*

Closed-ended questions work well when the major answers to questions are well known. Another plus is that interviews based on closed-ended questions do not necessarily require a large time commitment—more topics can be covered. As opposed to collecting such information via questionnaires, you can see body language and hear voice tone which can aid in interpreting the interviewee's responses. Closed-ended questions can also be an easy way to begin an interview and to determine which line of open-ended questions to pursue. You can include an "other" option to encourage the interviewee to add unanticipated responses. A major disadvantage of closed-ended questions is that useful information that does not quite fit

Interviewee:	Date:
Questions:	Notes:
When to ask question, if conditional *Question number: 1* 　　*Have you used the current sales* *tracking system? If so, how often?* *If yes, go to Question 2*	*Answer* 　　*Yes, I ask for a report on my* *product line weekly* *Observations* 　　*Seemed anxious — may be* *over-estimating usage frequency*
Question: 2 　　*What do you like least about this* *system?*	*Answer* 　　*Sales are shown in units, not* *dollars* *Observations* 　　*System can show sales in dollars,* *but user does not know this.*

FIGURE 2-6 *(Continued)   Typical Interview Guide*

into the defined answers may be overlooked as the respondent tries to make a choice instead of providing his or her best answer.

Closed-ended questions, like objective questions on an examination, can follow several forms, including these choices:

- True or false.
- Multiple choice (with only one response or selecting all relevant choices).
- Rating a response or idea on some scale, say from bad to good or strongly agree to strongly disagree. Each point on the scale should have a clear and consistent meaning to each person and there is usually a neutral point in the middle of the scale.
- Ranking items in order of importance.

INTERVIEW GUIDELINES   First, with either open- or closed-ended questions, do not phrase a question in a way that implies a right or wrong answer. The respondent must feel that he or she can state his or her true opinion and perspective and that his or her idea will be considered equally with those of others. Questions such as, "Should the system continue to provide the ability to override the default value, even though most users now do not like the feature?" should be avoided as such wording predefines a socially acceptable answer.

The second guideline to remember about interviews is to listen very carefully to what is being said. Take careful notes or, if possible, record the interview on a tape recorder (be sure to ask permission first!). The answers may contain extremely important information for the project. Also, this may be the only chance you have to get information from this particular person. If you run out of time and still need to get information from the person you are talking to, ask to schedule a follow-up interview.

Third, once the interview is over, go back to your office and type up your notes within 48 hours. If you recorded the interview, use the recording to verify the material in your notes. After 48 hours, your memory of the interview will fade quickly. As you type and organize your notes, write down any additional questions that might arise from lapses in your notes or from ambiguous information. Separate facts from your opinions and interpretations. Make a list of unclear points that need clarification. Call the person you interviewed and get answers to these new questions. Use the phone call as an opportunity to verify the accuracy of your notes. You may also want to send a written copy of your notes to the person you interviewed so the person can check your notes for accuracy. Finally, make sure you thank the person for his or her time. You may need to talk to your respondent again. If the interviewee will be a user of your system or is involved in some other way in the system's success, you want to leave a good impression.

Fourth, be careful during the interview not to set expectations about the new or replacement system unless you are sure these features will be part of the delivered system. Let the interviewee know that there are many steps to the project and the perspectives of many people need to be considered, along with what is technically possible. Let respondents know that their ideas will be carefully considered, but that due to the iterative nature of the systems development process, it is premature to say now exactly what the ultimate system will or will not do.

Fifth, seek a variety of perspectives from the interviews. Find out what potential users of the system, users of other systems that might be affected by changes, managers and superiors, information systems staff who have experience with the current system, and others think the current problems and opportunities are and what new information services might better serve the organization. You want to understand all possible perspectives so that in a later approval step you will have information on which to base a recommendation or design decision that all stakeholders can accept.

■ ADMINISTERING QUESTIONNAIRES

Interviews are very effective ways of communicating with people and obtaining important information from them. However, interviews are also very expensive and time-consuming to conduct. Thus, a limited number of questions can be covered and people contacted. In contrast, questionnaires are passive and often yield less depth of understanding than interviews; however, questionnaires are not as expensive to administer per respondent. In addition, questionnaires have the advantage of gathering information from many people in a relatively short time and of being less biased in the interpretation of their results.

CHOOSING QUESTIONNAIRE RESPONDENTS   Sometimes there are more people to survey than you can handle and you must decide which set of people to send the questionnaire to or which questionnaire to send to which group of people. Whichever group of respondents you choose, it should be representative of all users. In general, you can achieve a representative sample by any one or any combination of these four methods:

1. Those *convenient* to sample: these may be people at a local site, those willing to be surveyed, or those most motivated to respond

2. A *random* group: if you get a list of all users of the current system, simply choose every *n*th person on the list; or, you could select people by skipping names on the list based on numbers from a random number table

3. A *purposeful* sample: here you may specify only people who satisfy certain criteria, such as users of the system for more than two years or users who use the system most often

4. A *stratified* sample: in this case, you have several categories of people whom you definitely want to include—choose a random set from each category (e.g., users, managers, foreign business unit users)

Samples which combine characteristics of several approaches are also common. In any case, once the questionnaires are returned, you should check for non-response bias; that is, a systematic bias in the results since those who responded are different from those who did not respond. You can refer to books on survey research to find out how to determine if your results are confounded by non-response bias.

DESIGNING QUESTIONNAIRES   Questionnaires are usually administered on paper although they can be administered in person (resembling a structured interview), over the phone (computer-assisted telephone interviewing), or even on diskette. Questionnaires are less expensive, however, if they do not require a person to administer them directly; that is, if the people answering the questions can complete the questionnaire without help. Also, answers can be provided at the convenience of the respondent, as long as the answers are returned by a specific date.

Questionnaires typically include closed-ended questions, more than can be effectively asked in an interview, and sometimes contain open-ended questions as well. Closed-ended questions are preferable because they are easier to complete and they define the exact coverage required. A few open-ended questions give the person being surveyed an opportunity to add insights not anticipated by the designer of the questionnaire. In general, questionnaires take less time to complete than interviews structured to obtain the same information. In addition, questionnaires are given to many people simultaneously whereas interviews are usually limited to one person at a time.

Questionnaires are generally less rich in information content than interviews, however, because they provide no direct means by which to ask follow-up questions (although it is possible, though time-consuming, to contact respondents after they have returned their completed questionnaires to ask for further information). Also, since questionnaires are written, they do not provide the opportunity to judge the accuracy of the responses. In an interview, you can sometimes determine if people are answering truthfully or fully by the words they use, whether they make direct eye contact, the tone of voice they use, or their body language.

The ability to create good questionnaires is a skill that improves with practice and experience. Because the questions are written, they must be extremely clear in meaning and logical in sequence. When a person is completing a questionnaire, he or she only has the written questions to interpret and answer. You are not there to clarify each question's meaning. For example, what if a closed-ended question were phrased in this way:

How often do you back up your computer files?
a. Frequently
b. Sometimes
c. Hardly at all
d. Never

There are at least two sources of ambiguity in the wording of the question. The first source of ambiguity is the categories offered for the answer: the only non-ambiguous answer is "never." "Hardly at all" could mean anything from once per year to once per month, depending on who is answering the question. "Sometimes" could cover the same range of possibilities as "Hardly at all." "Frequently" could be anything from once per hour to once per week. The second source of ambiguity is in the question itself. Does the term "computer files" pertain only to those on my hard disk? Or does it also mean the files I have stored on floppy disk? What if I have more than one PC in my office? And what about the files I have stored on the minicomputer I use for certain applications? I don't back up those files; the system operator does it on a regular basis for all minicomputer files, not just mine. With no questioner present to explain the ambiguities, the respondent is at a loss and must try to answer the question in the best way he or she knows how. Whether the respondent's interpretation is the same as other respondents' is anyone's guess. The respondent cannot be there when the data are analyzed to tell exactly what was meant.

A less ambiguous way to phrase the question and its response categories would be something like this:

How often do you back up the computer files stored on the hard disk on the PC you use for over 50% of your work time?
a. Frequently (at least once per week)
b. Sometimes (from one to three times per month)
c. Hardly at all (once per month or less)
d. Never

As you can see from the wording of the question, the phrasing is a bit awkward, but it avoids ambiguity. You may want to break up a single question into multiple questions, or a set of questions and statements, to avoid awkward phrasing. Notice also that the possible responses are much clearer now that they have been specifically defined, and they cover the full range of possibilities, from never to at least once per week with no overlapping time periods.

Obviously, care must be taken in the task of composing closed-ended and open-ended questions. Further, you should be as careful in composing questions for interviews as for questionnaires, since sloppily worded questions cannot be identified every time in an interview unless the interviewee asks for clarification. For both interviews and questionnaires, it is wise to pretest your questions. Pose the questions in a simulated interview and ask the interviewee to rephrase each question as he or she interprets the question. Check responses for reasonableness. You can even ask the same question in what you think are several different ways to see if you receive a materially different response. Use this feedback to adjust the questions to make them less ambiguous.

Questionnaires are most useful in the requirements determination process when used for very specific purposes rather than for more general information gathering. For example, one useful application of questionnaires is to measure levels of user satisfaction with a system or with particular aspects of it. Another useful application is to have several users choose from among a list of system features available in many off-the-shelf software packages. You could ask users to choose the features they most want and quickly tabulate the results to find out which features are most in demand. You could then recommend a system solution based on a particular software package to meet the demands of most of the users.

◼ CHOOSING BETWEEN INTERVIEWS AND QUESTIONNAIRES

To summarize the previous sections, you can see that interviews are good tools for collecting rich, detailed information and that interviews allow exploration and follow-up (see Table 2.4). On the other hand, interviews are quite time-intensive and expensive. In comparison, questionnaires are inexpensive and take less time, as specific information can be gathered from many people at once without the personal intervention of an interviewer. The information collected from a questionnaire is less rich, however, and is potentially ambiguous if questions are not phrased precisely. In addition, follow-up to a questionnaire is more difficult as it often involves interviews or phone calls, adding to the expense of the process.

These differences and others are important for you to remember during the analysis phase. Deciding which method to use and what strategy to employ to gather information will vary with the system being studied and its organizational context. For example, if the organization is large and the system being studied is vast and complex, then there will probably

TABLE 2.4
Comparison of Interviews and Questionnaires

Characteristic	Interviews	Questionnaires
Information Richness	High (many channels)	Medium to low (only responses)
Time Required	Can be extensive	Low to moderate
Expense	Can be high	Moderate
Chance for Follow-up and Probing	Good: probing and clarification questions can be asked by either	Limited: probing and follow-up done after original data collection interviewer or interviewee
Confidentiality	Interviewee is known to interviewer	Respondent can be unknown
Involvement of Subject	Interviewee is involved and committed	Respondent is passive, no clear commitment
Potential Audience	Limited numbers, but complete responses from those interviewed	Can be quite large, but lack of response from some can bias results

be dozens of affected users and stakeholders. If you know little about the system or the organization, a good strategy is to identify key users and stakeholders and interview them. You would then use the information gathered in the interviews to create a questionnaire which would be distributed to a large number of users. You could then schedule follow-up interviews with a few users. At the other extreme, if the system and organization are small and you understand them well, the best strategy may be to interview only one or two key users or stakeholders.

■ INTERVIEWING GROUPS

One drawback to using interviews and questionnaires to collect systems requirements is the need for the analyst to reconcile apparent contradictions in the information collected. A series of interviews may turn up inconsistent information about the current system or its replacement. You must work through all of these inconsistencies to figure out what the most accurate representation of current and future systems might be. Such a process requires several follow-up phone calls and additional interviews. Catching important people in their offices is often difficult and frustrating, and scheduling new interviews may become very time-consuming. In addition, new interviews may reveal new questions that in turn require additional interviews with those interviewed earlier. Clearly, gathering information about an information system through a series of individual interviews and follow-up calls is not an efficient process.

Another option available to you is the group interview. In a group interview, you interview several key people at once. To make sure all of the important information is collected, you may conduct the interview with one or more analysts. In the case of multiple interviewers, one analyst may ask questions while another takes notes, or different analysts might concentrate on different kinds of information. For example, one analyst may listen for data requirements while another notes the timing and triggering of key events. The number of interviewees involved in the process may range from two to however many you believe can be comfortably accommodated.

A group interview has a few advantages. One, it is a much more effective use of your time than is a series of interviews with individuals (although the time commitment of the interviewees may be more of a concern). Two, interviewing several people together allows them to hear the opinions of other key people and gives them the opportunity to agree or disagree with their peers. Synergies also often occur. For example, the comments of one person might cause another person to say, "That reminds me of . . ." or "I didn't know that was a problem." You can benefit from such a discussion as it helps you identify issues on which there is general agreement and areas where views diverge widely.

The primary disadvantage of a group interview is the difficulty in scheduling it. The more people involved, the more difficult it will be finding a convenient time and place for everyone. Modern technology such as video conferences and video phones can minimize the geographical dispersion factors that make scheduling meetings so difficult. Group interviews are at the core of the Joint Application Design process, which we will discuss later.

■ DIRECTLY OBSERVING USERS

All the methods of collecting information that we have been discussing up until now involve getting people to recall and convey information they have about an organizational area and the information systems which support these processes. People, however, are not always very reliable informants, even when they try to be reliable and tell what they think is the truth. As odd as it may sound, people often do not have a completely accurate appreciation of what they do or how they do it. This is especially true concerning infrequent events, issues from the past, or issues for which people have considerable passion. Since people cannot always be trusted to reliably interpret and report their own actions, you can supplement and corroborate what people tell you by watching what they do or by obtaining relatively objective measures of how people behave in work situations. (See the box "Lost Soft Drink Sales" for an example of the importance of systems analysts learning firsthand about the business for which they are designing systems.)

For example, one possible view of how a hypothetical manager does her job is that a manager carefully plans her activities, works long and consistently on solving problems, and controls the pace of her work. A manager might tell you that is how she spends her day. When Mintzberg (1973) observed how managers work, however, he found that a manager's day is actually punctuated by many, many interruptions. Managers work in a fragmented manner, focusing on a problem or on a communication for only a short time before they are interrupted by phone calls or visits from their subordinates and other managers. An information system designed to fit the work environment described by our hypothetical manager would not effectively support the actual work environment in which that manager finds herself.

As another example, consider the difference between what another employee might tell you about how much he uses electronic mail and how much electronic mail use you might discover through more objective means. An employee might tell you he is swamped with e-mail messages and that he spends a significant proportion of his time responding to e-mail messages. However, if you were able to check electronic mail records, you might find that this employee receives only three e-mail messages per day on average, and that the most messages he has ever received during one eight-hour period is ten. In this case, you were able to obtain an accurate behavioral measure of how much e-mail this employee copes with without having to watch him read his e-mail.

The intent behind obtaining system records and direct observation is the same, however, and that is to obtain more firsthand and objective measures of employee interaction with information systems. In some cases, behavioral measures will be a more accurate reflection of reality than what employees themselves believe. In other cases, the behavioral information will substantiate what employees have told you directly. Although observation and obtaining objective measures are desirable ways to collect pertinent information, such methods are not always possible in real organizational settings. Thus, these methods are not totally unbiased, just as no other one data-gathering method is unbiased.

Lost Soft Drink Sales

A systems analyst was quite surprised to read that sales of all soft drink products were lower, instead of higher, after a new delivery truck routing system was installed. The software was designed to reduce stock-outs at customer sites by allowing drivers to visit each customer more often using more efficient delivery routes.

Confused by the results, management asked the analyst to delay a scheduled vacation, but he insisted that he could look afresh at the system only after a few overdue days of rest and relaxation.

Instead of taking a vacation, however, the analyst called a delivery dispatcher he had interviewed during the design of the system and asked to be given a route for a few days. The analyst drove a route (for a regular driver actually on vacation), following the schedule developed from the new system. What the analyst discovered was that the route was very efficient, as expected; so at first the analyst could not see any reason for lost sales.

During the third and last day of his "vacation" the analyst stayed overtime at one store to ask the manager if she had any ideas why sales might have dropped off in recent weeks. The manager had no explanation, but did make a seemingly unrelated observation that the regular route driver appeared to have less time to spend in the store. He did not seem to take as much interest in where the products were displayed and did not ask for promotional signs to be displayed, as he had often done in the past.

From this conversation, the analyst concluded that the new delivery truck routing system was, in one sense, too good. It placed the driver on such a tight schedule that a driver had no time left for the "schmoozing" required to get special treatment, that gave the company's products an edge over the competition.

Without first-hand observation of the system in action participating as a system user, the analyst might never have discovered the true problem with the system design. Once time was allotted for not only stocking new products but also for necessary marketing work, product sales returned to and exceeded levels achieved before the new system had been introduced.

For example, observation can cause people to change their normal operating behavior. Employees who know they are being observed may be nervous and make more mistakes than normal, may be careful to follow exact procedures which they do not typically follow, and may work faster or slower than normal. Moreover, since observation typically cannot be continuous, you receive only a snapshot image of the person or task you observe which may not include important events or activities. Since observation is very time-consuming, you will not only observe for a limited time but also a limited number of people and at a limited number of sites. Again, observation yields only a small segment of data from a possibly vast variety of data sources. Exactly which people or sites to observe is a difficult selection problem. You want to pick both typical and atypical people and sites and observe during normal and abnormal conditions and times to receive the richest possible data from observation.

■ ANALYZING PROCEDURES AND OTHER DOCUMENTS

As noted above, asking questions of the people who use a system every day or who have an interest in a system is an effective way to gather information about current and future systems. Observing current system users is a more direct way of seeing how an existing system operates, but even this method provides limited exposure to all aspects of current operations. These methods of determining system requirements can be enhanced by examining system and organizational documentation to discover more details about current systems and the organization these systems support.

Although we discuss here several important types of documents that are useful in understanding possible future system requirements, our discussion does not exhaust all possibilities. You should attempt to find all written documents about the organizational areas relevant to the systems under redesign. Besides the few specific documents we discuss, organizational mission statements, business plans, organization charts, business policy manuals, job descriptions, internal and external correspondence, and reports from prior organizational studies can all provide valuable insight.

What can the analysis of documents tell you about the requirements for a new system? In documents you can find information about

- Problems with existing systems (e.g., missing information or redundant steps)

- Opportunities to meet new needs if only certain information or information processing were available (e.g., analysis of sales based on customer type)

- Organizational direction that can influence information system requirements (e.g., trying to link customers and suppliers more closely to the organization)

- Titles and names of key individuals who have an interest in relevant existing systems (e.g., the name of a sales manager who led a study of buying behavior of key customers)

- Values of the organization or individuals who can help determine priorities for different capabilities desired by different users (e.g., maintaining market share even if it means lower short-term profits)

- Special information processing circumstances that occur irregularly that may not be identified by any other requirements determination technique (e.g., special handling needed for a few very large-volume customers and which requires use of customized customer ordering procedures)

- The reason why current systems are designed as they are, which can suggest features left out of current software which may now be feasible and more desirable (e.g., data about a customer's purchase of competitors' products were not available when the current system was designed; these data are now available from several sources)

- Data, rules for processing data, and principles by which the organization operates that must be enforced by the information system (e.g., each customer is assigned exactly one sales department staff member as a primary contact if the customer has any questions)

One type of useful document is a written work procedure for an individual or a work group. The procedure describes how a particular job or task is performed, including data and information that are used and created in the process of performing the job. For example, the procedure shown in Figure 2-7 includes data (list of features and advantages, drawings, inventor name, and witness names) required to prepare an invention disclosure. It also indicates that besides the inventor, the vice president for research and department head and dean must review the material, and that a witness is required for any filing of an invention disclosure. These insights clearly affect what data must be kept, to whom information must be sent, and the rules that govern valid forms.

GUIDE FOR PREPARATION OF INVENTION DISCLOSURE
(See FACULTY and STAFF MANUALS for detailed Patent Policy and routing procedures.)

(1) DISCLOSE ONLY ONE INVENTION PER FORM.

(2) PREPARE COMPLETE DISCLOSURE.

The disclosure of your invention is adequate for patent purposes ONLY if it enables a person skilled in the art to understand the invention.

(3) CONSIDER THE FOLLOWING IN PREPARING A COMPLETE DISCLOSURE:

(a) All essential elements of the invention, their relationship to one another, and their mode of operation.

(b) Equivalents that can be substituted for any elements.

(c) List of features believed to be new.

(d) Advantages this invention has over the prior art.

(e) Whether the invention has been built and/or tested.

(4) PROVIDE APPROPRIATE ADDITIONAL MATERIAL.

Drawings and descriptive material should be provided as needed to clarify the disclosure. Each page of this material must be signed and dated by each inventor and properly witnessed. A copy of any current and/or planned publication relating to the invention should be included.

(5) INDICATE PRIOR KNOWLEDGE AND INFORMATION.

Pertinent publications, patents or previous devices, and related research or engineering activities should be identified.

(6) HAVE DISCLOSURE WITNESSED.

Persons other than co-inventors should serve as witnesses and should sign each sheet of the disclosure only after reading and understanding the disclosure.

(7) FORWARD ORIGINAL PLUS ONE COPY (two copies if supported by grant/contract) TO VICE PRESIDENT FOR RESEARCH VIA DEPARTMENT HEAD AND DEAN.

FIGURE 2-7   *Example of a Procedure*

Procedures are not trouble-free sources of information, however. Sometimes your analysis of several written procedures will reveal a duplication of effort in two or more jobs. You should call such duplication to the attention of management as an issue to be resolved before system design can proceed. That is, it may be necessary to redesign the organization before the redesign of an information system can achieve its full benefits. Another problem you may encounter with a procedure occurs when the procedure is missing. Again, it is not your job to create a document for a missing procedure—that is up to management. A third and common problem with a written procedure happens when the procedure is out of date. You may realize the procedure is out of date when you interview the person responsible for performing the task described in the procedure. Once again, the decision to rewrite the procedure so that it matches reality is made by management, but you may make suggestions based upon your understanding of the organization. A fourth problem often encountered with written procedures is that the formal procedures may contradict information you collected from interviews, questionnaires, and observation about how the organization operates and what information is required. As in the other cases, resolution rests with management.

All of these problems illustrate the difference between formal systems and informal systems. Formal systems are systems recognized by the official documentation of the organization; informal systems are the way in which the organization actually works. Informal systems develop because of inadequacies of formal procedures, individual work habits and preferences, resistance to control, and other factors. It is important to understand both formal and informal systems since each provides insight into information requirements and what will be required to convert from present to future information services.

A second type of document useful to systems analysts is a business form (see a mock-up of a form in Figure 2-8). Forms are used for all types of business functions, from recording an order to acknowledging the payment of a bill to indicating what goods have been shipped. Forms are important for understanding a system because they explicitly indicate what data flow in or out of a system and which are necessary for the system to function. In the sample invoice form in Figure 2-8, we see data such as the name of the customer, the customer's sold to and ship to addresses, method of payment, data (item number, quantity, etc.) about each line item on the invoice, and calculated data such as tax and totals.

The form gives us crucial information about the nature of the organization. For example, the company can ship and bill to different addresses; item numbers appear to be all-numeric and five digits long; and the freight expense is charged to the customer. A printed form may correspond to a computer display that the system will generate for someone to

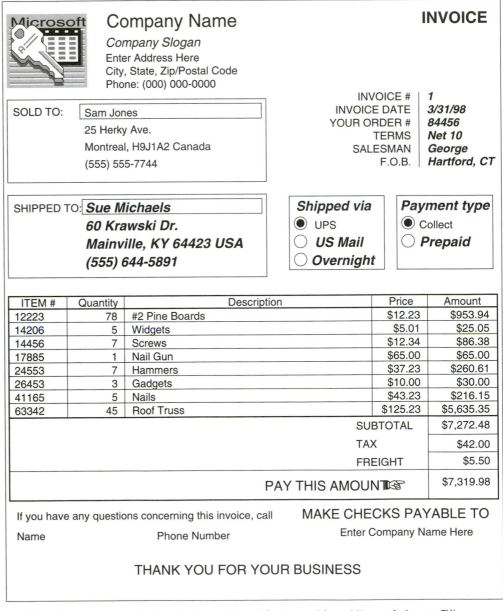

FIGURE 2-8   *Example of a Business Form (Generated from Microsoft Access™)*

enter and maintain data or to display data to on-line users. Forms are most useful to you when they contain actual organizational data (as in Figure 2-8), as this allows you to determine the characteristics of data which are actually used by the application. The ways in which people use forms change over time, and data that were needed when a form was designed may no longer be required. You can use the systems analysis techniques to help you determine which data are no longer required.

A third type of useful document is a report generated by current systems. As the primary output for some types of systems, a report enables you to work backwards from the information on the report to the data that must have been necessary to generate them. Figure 2-9 presents an example of a typical report prepared for management in a video rental business. This report shows that the system must be able to identify categories for each video and associate customers with the videos they rent. The date near the top of the report is probably the date on which the report was prepared and need not be kept by the system. You would analyze such reports to determine which data need to be captured over what time period and what manipulation of these raw data would be necessary to produce each field on the report.

If the current system is computer-based, a fourth set of useful documents are those that describe the current information systems—how they were designed and how they work. There are a lot of different types of documents that fit this description, everything from flow charts to data dictionaries and CASE tool reports to user manuals. An analyst who has access to such documents is lucky, as many in-house–developed information systems lack complete documentation (unless a CASE tool has been used).

Who rents what category of videos?

28-Feb-98   1

Video Category	Customer Name
action	
	Eleanor Johnson
	Miles Standish
foreign	
	Alan Alda
	Bob Hope
	Eleanor Johnson
	Jane Smith
	Juan Valdez
sf	
	Alan Alda
	Alexi Kosygin
	Bob Hope
	Eleanor Johnson
	Jane Smith
	John Smith
	Wanda Orlikowski
western	
	Alexi Kosygin
	Bob Hope
	Jane Smith
	Juan Valdez
	Miles Standish
	Wilma Randolph

FIGURE 2-9   *Example of a Report*

Analysis of organizational documents and observation, along with interviewing and questionnaires, are the methods most used for gathering system requirements. In Table 2.4 we summarized the comparative features of interviews and questionnaires. Table 2.5 summarizes the comparative features of observation and analysis of organizational documents.

TABLE 2.5
Comparison of Observation and Document Analysis

Characteristic	Observation	Document Analysis
Information Richness	High (many channels)	Low (passive) and old
Time Required	Can be extensive	Low to moderate
Expense	Can be high	Low to moderate
Chance for Follow-up and Probing	Good: probing and clarification questions can be asked during or after observation	Limited: probing possible only if original author is available
Confidentiality	Observee is known to interviewer; observee may change behavior when observed	Depends on nature of document; does not change simply by being read
Involvement of Subject	Interviewees may or may not be involved and committed depending on whether they know if they are being observed	None, no clear commitment
Potential Audience	Limited numbers and limited time (snapshot) of each	Potentially biased by which documents were kept or because document not created for this purpose

MODERN METHODS FOR DETERMINING SYSTEM REQUIREMENTS

Even though we called interviews, questionnaires, observation, and document analysis traditional methods for determining a system's requirements, all of these methods are still very much used by analysts to collect important information. Today, however, there are additional techniques to collect information about the current system, the organizational area requesting the new system, and what the new system should be like. In this section, you will learn about several modern information-gathering techniques for analysis (listed in Table 2.6): Joint Application Design (JAD), group support systems, CASE tools, and prototyping. As we said earlier, these techniques can support effective information collection and structuring while reducing the amount of time required for analysis. There is an alternative to SDLC called RAD, which combines JAD, CASE tools, and prototyping.

TABLE 2.6
Modern Methods for Collecting System Requirements

- Bringing together in a *Joint Application Design (JAD)* session users, sponsors, analysts and others to discuss and review system requirements
- Using *group support systems* to facilitate the sharing of ideas and voicing opinions about system requirements
- Using *CASE* tools to analyze current systems to discover requirements to meet changing business conditions
- Iteratively developing system *prototypes* that refine the understanding of system requirements in concrete by showing working versions of system features

■ JOINT APPLICATION DESIGN

JAD started in the late 1970s at IBM and since then the practice of JAD has spread throughout many companies and industries. For example, it is quite popular in the insurance industry in Connecticut where a JAD users group has been formed. In fact, several generic approaches to JAD have been documented and popularized (see Wood and Silver, 1989, for an example). The main idea behind JAD is to bring together the key users, managers, and systems analysts involved in the analysis of a current system. In that respect, JAD is similar to a group interview; a JAD, however, follows a particular structure of roles and agenda that is quite different from a group interview during which analysts control the sequence of questions answered by users. The primary purpose of using JAD in the analysis phase is to collect systems requirements simultaneously from the key people involved with the system. The result is an intense and structured, but highly effective, process. As with a group interview, having all the key people together in one place at one time allows analysts to see where there are areas of agreement and where there are conflicts. Meeting with all these important people for over a week of intense sessions allows you the opportunity to resolve conflicts, or at least to understand why a conflict may not be simple to resolve.

JAD sessions are usually conducted in a location other than the place where the people involved normally work. The idea behind such a practice is to keep participants away from as many distractions as possible so that they can concentrate on systems analysis. A JAD may last anywhere from four hours to an entire week and may consist of several sessions. A JAD employs thousands of dollars of corporate resources, the most expensive of which is the time of the people involved. Other expenses include the costs associated with flying people to a remote site and putting them up in hotels and feeding them for several days.

The typical participants in a JAD are listed below:

- *JAD session leader:* The JAD leader organizes and runs the JAD. This person has been trained in group management and facilitation as well as in systems analysis. The JAD leader sets the agenda and sees that it is met. The JAD leader remains neutral on issues and does not contribute ideas or opinions but rather concentrates on keeping the group on the agenda, resolving conflicts and disagreements, and soliciting all ideas.

- *Users:* The key users of the system under consideration are vital participants in a JAD. They are the only ones who have a clear understanding of what it means to use the system on a daily basis.

- *Managers:* Managers of the work groups who use the system in question provide insight into new organizational directions, motivations for and organizational impacts of systems, and support for requirements determined during the JAD.

- *Sponsor:* As a major undertaking due to its expense, a JAD must be sponsored by someone at a relatively high level in the company. If the sponsor attends any sessions, it is usually only at the very beginning or the end.

- *Systems Analysts:* Members of the systems analysis team attend the JAD although their actual participation may be limited. Analysts are there to learn from users and managers, not to run or dominate the process.

- *Scribe:* The scribe takes notes during the JAD sessions. This is usually done on a personal computer or laptop. Notes may be taken using a word processor, or notes and diagrams may be entered directly into a CASE tool.

- *IS staff:* Besides systems analysts, other IS staff, such as programmers, database analysts, IS planners, and data center personnel, may attend to learn from the discussion and possibly to contribute their ideas on the technical feasibility of proposed ideas or on technical limitations of current systems.

JAD sessions are usually held in special-purpose rooms where participants sit around horseshoe-shaped tables, as in Figure 2-10. These rooms are typically equipped with whiteboards (possibly electronic, with a printer to make copies of what is written on the board). Other audio-visual tools may be used, such as transparencies and overhead projectors, magnetic symbols that can be easily rearranged on a whiteboard, flip charts, and computer-generated displays. Flip chart paper is typically used for keeping track of issues that cannot be resolved during the JAD or for those issues requiring additional information that can be gathered during breaks in the proceedings. Computers may be used to create and display form or report designs or for diagramming existing or replacement systems. In general, however, most JADs do not benefit much from computer support (Carmel, 1991).

When a JAD is completed, the end result is a set of documents that detail the workings of the current system related to the study of a replacement system. Depending on the exact purpose of the JAD, analysts may also walk away from the JAD with some detailed information on what is desired of the replacement system.

TAKING PART IN A JAD   Imagine that you are a systems analyst taking part in your first JAD. What might participating in a JAD be like? Typically, JADs are held off site, in comfortable conference facilities. On the first morning of the JAD, you and your fellow analysts walk into a room that looks much like the one depicted in Figure 2-6. The JAD facilitator is already there; she is finishing writing the day's agenda on a flip chart. The scribe is seated in a corner at a microcomputer, preparing to take notes on the day's activities. Users and managers begin to enter in groups and seat themselves around the U-shaped table. You and the other analysts review your notes describing what you have learned so far about the information system you are all here to discuss. The session leader opens the meeting with a welcome and a brief run-down of the agenda. The first day will be devoted to a general overview of the current system and major problems associated with it. The next two days will be devoted to an analysis of current system screens. The last two days will be devoted to analysis of reports.

The session leader introduces the corporate sponsor who talks about the organizational unit and current system related to the systems analysis study and the importance of upgrading the current system to meet changing business conditions. He leaves, and the JAD session leader takes over. She yields the floor to the senior analyst who begins a presentation on key

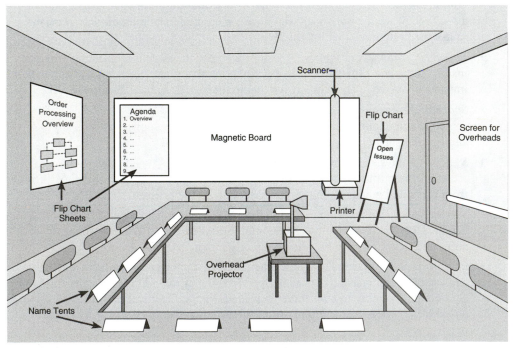

FIGURE 2-10   *Illustration of the Typical Room Layout for a JAD (Adapted from Wood and Silver, 1989)*

problems with the system that have already been identified. After the presentation, the session leader opens the discussion to the users and managers in the room.

After a few minutes of talk, a heated discussion begins between two users from different corporate locations. One user, who represents the office which served as the model for the original systems design, argues that the system's perceived lack of flexibility is really an asset, not a problem. The other user, who represents an office that was part of another company before a merger, argues that the current system is so inflexible as to be virtually unusable. The session leader intervenes and tries to help the users isolate particular aspects of the system that may contribute to the system's perceived lack of flexibility.

Questions arise about the intent of the original developers. The session leader asks the analysis team about their impressions of the original system design. Because these questions cannot be answered during this meeting, as none of the original designers are present and none of the original design documents are readily available, the session leader assigns the question about intent to the "to do" list. This question becomes the first one on a flip chart sheet of "to do" items and the session leader gives you the assignment of finding out about the intent of the original designers. She writes your name next to the "to do" item on the list and continues with the session. Before the end of the JAD, you must get an answer to this question.

The JAD will continue like this for its duration. Analysts will make presentations, help lead discussions of form and report design, answer questions from users and managers, and take notes on what is being said. After each meeting, the analysis team will meet, usually informally, to discuss what has occurred that day and to consolidate what they have learned. Users will continue to contribute during the meetings and the session leader will facilitate, intervening in conflicts, seeing that the group follows the agenda. When the JAD is over, the session leader and her assistants must prepare a report that documents the findings in the JAD and is circulated among users and analysts.

CASE TOOLS DURING JAD The CASE tools most useful to analysis during a JAD are those referred to as upper CASE, as they apply most directly to activities occurring early in the systems development life cycle. Upper CASE tools usually include planning tools, diagramming tools, and prototyping tools, such as computer display and report generators. For requirements determination and structuring, the most useful CASE tools are for diagramming and for display and report generation. The more interaction analysts have with users during this phase, the more useful this set of tools. The analyst can use diagramming and prototyping tools to give graphic form to system requirements, show the tools to users, and make changes based on the users' reactions. The same tools are very valuable for requirements structuring as well. Using common CASE tools during requirements determination and structuring makes the transition between these two sub-phases easier and reduces the time spent. In structuring, CASE tools which analyze requirements information for correctness, completeness, and consistency are also useful. Finally, for alternative generation and selection, diagramming and prototyping tools are key to presenting users with graphic illustrations of what the alternative systems will look like. Such a practice provides users and analysts with better information to select the most desirable alternative system.

Some observers advocate using CASE tools during JADs (Lucas, 1993). Running a CASE tool during a JAD allows analysts to enter system models directly into a CASE tool, providing consistency and reliability in the joint model-building process. The CASE tool captures system requirements in a more flexible and useful way than can a scribe or an analysis team making notes. Further, the CASE tool can be used to project menu, display, and report designs, so users can directly observe old and new designs and evaluate their usefulness for the analysis team. Some CASE tools are too slow for the real-time pace of most JADs, however, so the session leader may want analysts to enter information into the CASE tool after the day's meeting is over. Users and managers can be shown the results of their work the next morning.

SUPPORTING JAD WITH GSS The JAD process is typically not well supported by computing, despite suggestions to augment JADs with CASE tools and other computer-based aids. Most traditional JADs rely on one computer for the scribe to use and maybe another for displaying screen and report designs. Since JAD is a structured group process, JAD can benefit from the same computer-based support that can be applied to any group process. Group support systems (GSS) can be used to support group meetings. Here we will discuss how JAD can benefit from GSS use.

One disadvantage to a JAD session is that it suffers from many of the same problems as any group meeting. For example, the more people in a group, the less time there is for all of them to speak and state their views. Even if you assumed that they all spoke for an equal amount of time, no one would have much time to talk in a one-hour meeting for 12 people (only five minutes each!). The assumption about speaking equally points out a second problem with meetings—one or a few people always dominate the discussion. On the other hand, some people will say absolutely nothing. Whatever outcome the meeting produces tends to be tilted toward those who spoke the most during the meeting and others may not be fully committed to the conclusions reached. A third problem with group meetings is that some people are afraid to speak out for fear they will be criticized. A fourth problem is that most people are not willing to criticize or challenge their bosses in a meeting, even if what the boss is saying is wrong.

JADs suffer from all of these problems with meetings. The result is that important views often are not aired. Such an outcome is unfortunate as the design of the new system could be adversely impacted and the system may have to be reworked at great expense when those important views finally become known.

GSSs have been designed specifically to help alleviate some of the problems with group meetings. In order to provide everyone in the meeting with the same chance to contribute, group members type their comments into computers rather than speak them. The GSS is set up so that all members of the group can see what every other member has been typing. In the one-hour meeting for 12 people mentioned earlier, all 12 can contribute for the full hour, instead of just for five minutes, using a GSS. If everyone in the meeting is typing, not talking, and everyone has the same chance to contribute, then the chances of domination of the meeting by any one individual are greatly reduced. Also, comments typed into a GSS are anonymous. Anonymity helps those who fear criticism because only the comment, and not the person, can be criticized, since no one knows who typed what. Anonymity also provides the ability to criticize your boss.

Supporting a JAD with a GSS has many potential benefits. Using a GSS, a JAD session leader is more likely to obtain contributions from everyone, rather than from just a few. Important ideas are less likely to be missed. Similarly, poor ideas are more likely to be criticized. A study comparing traditional JAD to JAD supported with GSS found that using a GSS did lead to certain enhancements in the JAD process (Carmel, George, and Nunamaker, 1992). Among the findings were that GSS-supported JADs tended to be more time-efficient than traditional JAD and participation was more equal because there was less domination by certain individuals than in traditional JAD. The study also found that introducing a GSS into a JAD session had other, less desirable, effects. GSS-supported JADs tended to be less structured and it was more difficult to identify and resolve conflicts when a GSS was used, due in part to the anonymity of interaction. Supporting a JAD with GSS, then, does seem to provide some benefits through altering how the group works together. Yet a reduction in the JAD leader's ability to resolve conflicts could be a problem, especially since JAD was designed to help uncover and resolve conflicts.

■ USING PROTOTYPING DURING REQUIREMENTS DETERMINATION

You were introduced to prototyping earlier. Prototyping is an iterative process involving analysts and users whereby a rudimentary version of an information system is built and rebuilt according to user feedback. Prototyping can replace the systems development life cycle or augment it. What we are interested in here is how prototyping can augment the requirements determination process.

In order to gather an initial basic set of requirements, you will still have to interview users and collect documentation. Prototyping, however, will allow you to quickly convert basic requirements into a working, though limited, version of the desired information system. The prototype will then be viewed and tested by the user. Typically, seeing verbal descriptions of requirements converted into a physical system will prompt the user to modify existing requirements and generate new ones. For example, in the initial interviews, a user might have said that he wanted all relevant utility billing information on a single computer display form, such as the client's name and address, the service record, and payment history. Once the same user sees how crowded and confusing such a design would be in the prototype, he might change his mind and instead ask for the information to be organized on several screens, but with easy transitions from one screen to another. He might also be reminded of some important requirements (data, calculations, etc.) that had not surfaced during the initial interviews.

You would then redesign the prototype to incorporate the suggested changes. Once modified, users would again view and test the prototype. And, once again, you would incorporate their suggestions for change. Through such an iterative process, the chances are good that you will be able to better capture a system's requirements. The goal with using prototyping to support requirements determination is to develop concrete specifications for the ultimate system, not to build the ultimate system from prototyping.

Prototyping is possible with several 4GLs and with CASE tools. As we saw there, you can use CASE tools as part of a JAD to provide a type of limited prototyping with a group of users.

Prototyping is most useful for requirements determination when

- User requirements are not clear or well understood, which is often the case for totally new systems or systems that support decision making
- One or a few users and other stakeholders are involved with the system
- Possible designs are complex and require concrete form to fully evaluate
- Communication problems have existed in the past between users and analysts and both parties want to be sure that system requirements are as specific as possible
- Tools (such as form and report generators) and data are readily available to rapidly build working systems

Prototyping also has some drawbacks as a tool for requirements determination. These include

- A tendency to avoid creating formal documentation of system requirements which can then make the system more difficult to develop into a fully working system
- Prototypes can become very idiosyncratic to the initial user and difficult to diffuse or adapt to other potential users
- Prototypes are often built as stand-alone systems, thus ignoring issues of sharing data and interactions with other existing systems
- Checks in the SDLC are bypassed so that some more subtle, but still important, system requirements might be forgotten (e.g., security, some data entry controls, or standardization of data across systems)

RADICAL METHODS FOR DETERMINING SYSTEM REQUIREMENTS

Whether traditional or modern, the methods for determining system requirements that you have read about in this section apply to any requirements determination effort, regardless of its motivation. But most of what you have learned has traditionally been applied to systems development projects that involve automating existing processes. Analysts use system requirements determination to understand current problems and opportunities, as well as what is needed and desired in future systems. Typically, the current way of doing things has a large impact on the new system. In some organizations, though, management is looking for new ways to perform current tasks. These new ways may be radically different from how things are done now, but the payoffs may be enormous: fewer people may be needed to do the same work, relationships with customers may improve dramatically, and processes may become much more efficient and effective, all of which can result in increased profits. The overall process by which current methods are replaced with radically new methods is generally referred to as business process reengineering or BPR.

To better understand BPR, consider the following analogy. Suppose you are a successful European golfer who has tuned your game to fit the style of golf courses and weather in Europe. You have learned how to control the flight of the ball in heavy winds, roll the ball on wide open greens, putt on large and undulating greens, and aim at a target without the aid of the landscaping common on North American courses. When you come to the United States to make your fortune on the U.S. tour, you discover that incrementally improving your putting, driving accuracy, and sand shots will help, but the new competitive environment is simply not suited to your style of the game. You need to reengineer your whole approach, learning how to aim at targets, spin and stop a ball on the green, and manage the distractions of crowds and press. If you are good enough, you may survive, but without reengineering, you will never be a winner.

Just as the competitiveness of golf forces good players to adapt their games to changing conditions, the competitiveness of our global economy has driven most companies into a mode of continuously improving the quality of their products and services (Dobyns and Crawford-Mason, 1991). Organizations realize that creatively using information technologies can yield significant improvements in most business processes. The idea behind BPR is not just to improve each business process but, in a systems modeling sense, to reorganize the complete flow of data in major sections of an organization to eliminate unnecessary steps, achieve synergies among previously separate steps, and become more responsive to future changes. Companies such as IBM, Procter & Gamble, Wal-Mart, and Ford are actively pursuing BPR efforts and have had great success. Yet, many other companies have found difficulty in applying BPR principles (Moad, 1994). Nonetheless, BPR concepts are actively applied in both corporate strategic planning and information systems planning as a way to radically improve business processes.

BPR advocates suggest that radical increases in the quality of business processes can be achieved through creative application of information technologies. BPR advocates also suggest that radical improvement cannot be achieved by tweaking existing processes but rather by using a clean sheet of paper and asking "If we were a new organization, how would we accomplish this activity?" Changing the way work is performed also changes the way information is shared and stored, which means that the results of many BPR efforts are the development of information system maintenance requests or requests for system replacement. It is likely that you will encounter or have encountered BPR initiatives in your own organization. A recent survey of IS executives found that they view BPR to be a top IS priority for the coming years (Hayley, Plewa, and Watts, 1993).

◼ IDENTIFYING PROCESSES TO REENGINEER

A first step in any BPR effort relates to understanding what processes to change. To do this, you must first understand which processes represent the key business processes for the organization. Key business processes are the structured set of measurable activities designed to produce a specific output for a particular customer or market. The important aspect of this definition is that key processes are focused on some type of organizational outcome such as the creation of a product or the delivery of a service. Key business processes are also customer-focused. In other words, key business processes would include all activities used to design, build, deliver, support, and service a particular product for a particular customer. BPR efforts, therefore, first try to understand those activities that are part of the organization's key business processes and then alter the sequence and structure of activities to achieve radical improvements in speed, quality, and customer satisfaction. The same techniques you learned to use for systems requirement determination can be used to discover and understand key business processes. Interviewing key individuals, observing activities, reading and studying organizational documents, and conducting JADs can all be used to find and fathom key business processes.

After identifying key business processes, the next step is to identify specific activities that can be radically improved through reengineering. Hammer and Champy (1993), the two people most identified with the term BPR, suggest that three questions be asked to identify activities for radical change:

1. How important is the activity to delivering an outcome?
2. How feasible is changing the activity?
3. How dysfunctional is the activity?

The answers to these questions provide guidance for selecting which activities to change. Those activities deemed important, changeable, yet dysfunctional, are primary candidates. To identify dysfunctional activities, they suggest you look for activities where there are excessive information exchanges between individuals, where information is redundantly recorded or needs to be rekeyed, where there are excessive inventory buffers or inspections, and where there is a lot of rework or complexity. Many of the tools and techniques for modeling data, processes, events, and logic within the IS development process are also being applied to model business processes within BPR efforts (see Davenport, 1993). Thus, the skills of a systems analyst are often central to many BPR efforts.

■ DISRUPTIVE TECHNOLOGIES

Once key business processes and activities have been identified, information technologies must be applied to radically improve business processes. To do this, Hammer and Champy (1993) suggest that organizations think "inductively" about information technology. Induction is the process of reasoning from the specific to the general, which means that managers must *learn* the power of new technologies and *think* of innovative ways to alter the way work is done. This is contrary to deductive thinking where problems are first identified and solutions are then formulated.

Hammer and Champy suggest that managers especially consider disruptive technologies when applying deductive thinking. Disruptive technologies are those that enable the breaking of long-held business rules that inhibit organizations from making radical business changes. For example, Saturn is using production schedule databases and electronic data interchange (EDI) to work with its suppliers as if they and Saturn were one company. Suppliers do not wait until Saturn sends them a purchase order for more parts but simply monitor inventory levels and automatically send shipments as needed (Hammer and Champy, 1993: 90). Table 2.7 shows several long-held business rules and beliefs that constrain organizations from making radical process improvements. For example, the first rule suggests that information can only appear in one place at a time. However, the advent of distributed databases has "disrupted" this long-held business belief.

BPR is increasingly being used to identify ways to adapt existing information systems to changing organizational information needs and processes. The specific tools and techniques for performing BPR are still evolving. For more information on this exciting topic, the interested reader is encouraged to see the books by Davenport (1993) and Hammer and Champy (1993).

TABLE 2.7
Long-Held Organizational Rules that Are
Being Eliminated Through Disruptive Technologies

Rule	Disruptive Technology
Information can appear in only one place at a time.	Distributed databases allow the sharing of information.
Only experts can perform complex work.	Expert systems can aid nonexperts.
Businesses must choose between centralization and decentralization.	Advanced telecommunications networks can support dynamic organizational structures.
Managers must make all decisions.	Decision-support tools can aid nonmanagers.
Field personnel need offices where they can receive, store, retrieve, and transmit information.	Wireless data communication and portable computers provide a "virtual" office for workers.
The best contact with a potential buyer is personal contact.	Interactive communication technologies allow complex messaging capabilities.
You have to find out where things are.	Automatic identification and tracking technology know where things are.
Plans get revised periodically.	High-performance computing can provide real-time updating.

SYSTEMS ANALYSIS

The purpose of systems analysis is to ascertain what must be done in order to carry out the functions of the system. This will involve a decomposition of the functions of the system into their logical constituents and the production of a logical model of the processes and of the data flows necessary to perform these. The logical model will be illustrated by data flow diagrams at the various levels of process decomposition. The algorithms will be revealed by structured process specification techniques such as structured English, decision tables and logic flow-charts. The importance of concentration on the logical features of a system (what logically needs to be done in order to carry out the functions) as distinct from the physical features (who does what, where, with which file and so on) is to avoid making premature commitments to physical design.

Prior to production of the logical analysis it is often helpful to carry out a physical analysis of the document flows between processes and departments within the existing system. As well as enabling the analyst to identify key tasks these charts can be used to evaluate control and efficiency aspects of the current system.

The output of the stage of systems analysis will be a logical model of the functioning of the system. This will consist of diagrams, charts and dictionaries, which are the product of the techniques used in the analysis. An important feature of a structured approach to systems analysis and design is the generation of clear and helpful documentation that can assist communication not only between the programmer and the analyst. The fact that a logical model is produced in systems analysis removes complicating and distracting physical elements that would hamper communication. The movement from the physical to the logical model and the techniques used are illustrated in Figure 2-11.

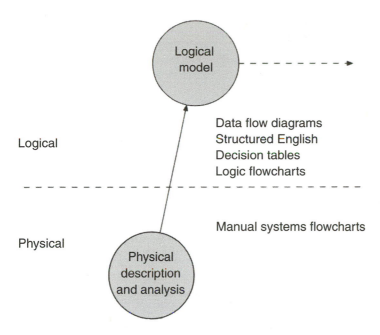

FIGURE 2-11   *Tools Used During the Stage of Systems Analysis*

MANUAL SYSTEMS FLOWCHARTS

After investigation the analyst may have collected an unwieldy batch of interview notes, details of observations, questionnaire responses and sundry documents. In the initial stages of analysis it is important to arrive at a methodical description of the existing manual system and to carry out some analysis at a physical level, prior to developing a logical model of the system. The flow of *formal* information within a system often occurs through documents that pass from one department to another. A traditional tool of systems analysis is the manual systems (document) flowchart.

The basic idea is that certain tasks performed on documents are common to many applications—filing, preparing multiple copies, collating, sorting. These are given special agreed symbols (Figure 2-12). The life history of a document from origination (entry from outside the system or preparation within) to destination (exit from the system or filing) is recorded on the flowchart. The passage of the document from department to department is also shown.

The best way to understand a manual systems flowchart, sometimes called a document flowchart, is to study one. Here the Kismet case study is developed giving a detailed description of the processes occurring during order processing. A manual systems flowchart covering these is shown in Figure 2-13.

■ PROCEDURES ADOPTED DURING ORDER PROCESSING

The customers mail their orders to Kismet HQ. On receipt of an order in the sales order department a five-part company order form is filled out giving (amongst other information) the **order#, order date, customer#, customer name, item 1 code#, item 1 quantity number, item 2 code#, item 2 quantity number** *and so on. The top copy of this form is temporarily filed in* **customer#** *sequence for customer enquiry purposes (rather than* **order#** *sequence, as customers enquire about a recently placed order they will not be in possession of the* **order#***). Each item is provisionally priced on the remaining copies of the form from a sales catalog held in the order department. The priced copies are sent to the credit control section.*

The credit control section provisionally calculates the order value. Brief details of the customer account are then consulted to establish that the customer exists, is correctly named, and that the total value of the order, when added to the current balance, does not exceed the cred-

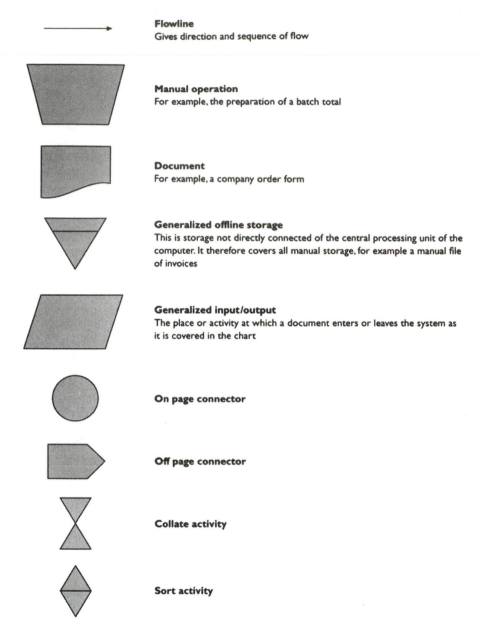

Flowline
Gives direction and sequence of flow

Manual operation
For example, the preparation of a batch total

Document
For example, a company order form

Generalized offline storage
This is storage not directly connected of the central processing unit of the computer. It therefore covers all manual storage, for example a manual file of invoices

Generalized input/output
The place or activity at which a document enters or leaves the system as it is covered in the chart

On page connector

Off page connector

Collate activity

Sort activity

FIGURE 2-12   *Basic Symbols Used in the Preparation of Manual Systems Flowcharts*

it limit of the customer. If all these conditions are met then the order copies are stamped 'approved,' signed, and returned to the sales order department, one copy of the order being retained in the credit control department filed by **customer#.** *If the above conditions are not met the order copies are temporarily filed to be dealt with later by the credit control manager.*

On receipt of the approved order copies in the sales order department, the top copy is extracted from the temporary file and sent to the customer as an acknowledgment. One of the 'approved' copies is filed in the order department in the approved order file under **order#.** *This is to enable staff to retrieve details of the order in the case of further customer queries. The remaining two copies are sent to the stores department and the invoicing department. The invoicing department files the copy under* **order#.**

The stores department selects the goods as ordered and enters the quantities supplied on the order form. A two-part despatch note is made out giving the goods supplied together with their quantities. One copy of this is sent to the invoicing department and the other is sent with the goods to packing and despatch. If the entire order is supplied then the order form is filed

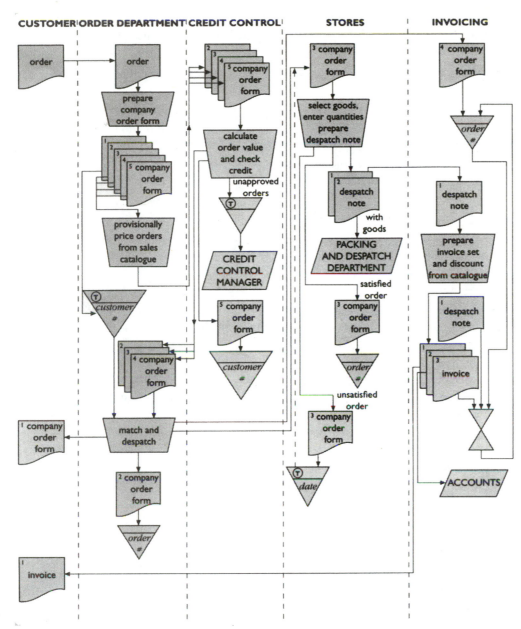

FIGURE 2-13 *The Manual System Flowchart for Order Processing/Despatch in Kismet*

in the stores department under **order#,** *otherwise the goods supplied are noted on the form and it is filed in* **date** *sequence. Periodically the stores department goes through the unsatisfied back orders and attempts to supply the goods ordered. The stores department also updates the inventory records.*

On receipt of the despatch note the invoicing department prepares a three-part invoice using the sales price of the goods from the catalog. The discount applicable to a customer is calculated. This is based on the customer's geographical location, total purchases during the last 12 months and the total value of the order. Sales tax is added and totals formed. One copy of the invoice is sent to the customer and one is sent to accounts for updating the customer accounts and other ledgers. The remaining copy is filed in the invoicing department with the order copy and despatch note under **order#.**

The flowchart for the order processing and despatch in Kismet is given in Figure 2-13. Note that the flow lines indicate flows of *documents*. The following practical points will assist in the drawing of flowcharts.

- The chart is divided into vertical sections representing different locations for operations.
- Though not shown, a far-left section may be used for additional (brief) narrative.
- The chart proceeds as far as possible from left to right and top to bottom.
- Documents are shown at origination, on entry into a section and then again only as required to avoid confusion with other documents.
- Ensure that all documents are accounted for by being permanently filed, being destroyed, leaving the system as charted (for example to the credit control manager in Kismet), or transferring to another flowchart.

There are a number of advantages and disadvantages in the use of manual systems flowcharts.

ADVANTAGES

- Flowcharts are easier to understand and assimilate than narrative. This becomes more pronounced with increasing complexity of the system.
- The preparation of a chart necessitates the full understanding by the analyst of the procedures and sequences of operations on documents.
- Incompleteness in tracing the destination of a document is easily discovered, indicating the need for further investigation on the part of the analyst (in Kismet, where does the customer's original order go?).
- Little technical knowledge is required to appreciate the document and so it can be used as a communication tool between the user of the system and the analyst in order to check and correct the latter's understanding.
- Weaknesses within the system, such as preparation of unnecessary documents, lack of control, unnecessary duplication of work and bottlenecks are easily located.

DISADVANTAGES

- With heavily integrated systems flowcharts may become difficult to manage (large sheets of paper!). The use of off-page connectors and continuation is sometimes necessary but tends to reduce the visual impact and clarity of the chart.
- They are difficult to amend.
- It must be realized that when analyzing an existing system informal information is an important part. The flowchart does not incorporate any recognition of this.

The systems flowchart is not only of use to the analyst when carrying out the stages of analysis and design of a computerized information system. Management may use the flowchart to impose uniformity on groups of systems as the structure of the processes surrounding document handling are revealed. This may be necessary to ensure that, say, one branch of an organization handles order processing in the same way as another. The flowchart may be used as an aid in the preparation of internal audit and procedures manuals. In the former case it is possible to ensure that essential information is provided to management at the correct stage. Auditors may use the flowchart in a review of internal control as a guide to determining auditing procedures in the annual audit.

The task of evaluation of the system is often considered as part of analysis. As has been pointed out, flowcharts assist in this task. A typical approach to evaluation of the order and despatch system of Kismet would use the chart to answer a number of questions. Note how easy it is to answer the following typical list of questions by using the flowchart:

1. Can goods be despatched but not invoiced?
2. Can orders be received and not (completely) dealt with?
3. Can customers be invoiced for goods that are not despatched because of low stocks?
4. Can goods be despatched to customers who are not credit-worthy?
5. Can invoicing errors occur?
6. Can sales be invoiced but not recorded?

DATA FLOW DIAGRAMS

Although systems flowcharts provide a useful tool for analyzing a physical description they may impede the design process. This is because they draw attention to physical detail. It is important to realize that the systems analyst will be designing a system to *do* something. This 'something' can be specified by describing its logic and the actions on data. To concentrate on existing physical detail will obscure the functions of the system, will restrict the designer's creativity, and will cause premature commitments to physical design in the early stages of the project.

For instance, it is of little importance to a computer design that one copy of a Kismet company order form is temporarily filed in the order department, while four copies go to credit control where, after approval, one is filed, the remaining copies being returned to the order department, after which the first copy is sent to the customer.

If the whole procedure is to be computerized, including the order approval, the process will occur within the computer as an exchange of data between files or a database and programs. But again, to assume total computerization is to make a possibly premature physical design decision. It may be more effective to retain parts of the old manual system.

The point to realize is that the processes, the exchanges of data and the stores of data are important, not their particular physical representation, whether it be, for instance, a sequential file on tape, an indexed file on desk or a composition of two manual files in two separate locations.

Data flow diagrams assist in building a logical model of the system independent of physical commitments. They show the various flows of data between the processes that transform it. Data stores and the points at which data enters and leaves a system are also shown. The basic symbols used in drawing data flow diagrams are shown in Figure 2-14.

- *Data source* or *data sink*: The square indicates a source or sink for data and is a reflection of the ignorance as to what happens to the data prior to its emergence from the source or after its disappearance into the sink.

- *Data process*: The circle or rounded rectangle indicates a data process. In this, a brief meaningful identifier of the process is written. It is important to realize that only *data* processes occur in a data flow diagram. Physical processes are not mentioned. For instance, the fact that goods are selected from their stock locations to satisfy an order is not a data process but a material task. A data process is one that transforms only data. It may be that this process will be carried out by a computer program, a part of a computer program, a set of computer programs or manually. The data flow diagram is neutral on this point.

  The identifier used in the data process symbol should, ideally, be both meaningful and succinct. It is good practice to restrict identifiers to a concatenation of imperative verb and object. For example *process stock transaction* or *check credit status* are both acceptable. It is bad practice to try to describe the process. For example, it is a great temptation for the novice to 'name' a data process as *check*

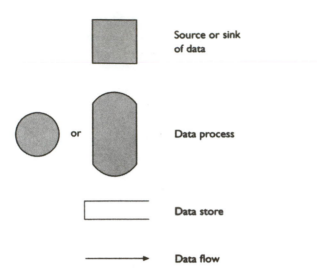

FIGURE 2-14   *Symbols used in Data Flow Diagrams*

the application for the account against the credit point list to establish the credit-worthiness and the credit limit of the customer. This is not acceptable.

- *Data store*: A data store is represented by an open-ended rectangle with a suitable identifier. This is distinguished from a data sink by the fact that the data is stored and it (or some part of it) will be retrieved. A sink indicates ignorance, as far as the diagram is concerned, of the fate of the data. No particular storage medium is implied by the data store symbol. It could be a magnetic tape or disk, a document or even some person's memory. Once again it is a great temptation for the newcomer to represent material stores. This is a mistake.

- *Data flow*: The line represents a data flow. The nature of the data is indicated by a name (or names). Wherever that piece of data flows within the diagram it should be tagged with the same name. Once again it is important to realize that these are flows of *data* and not material flows. If a data flow diagram is drawn representing part of a system in which goods accompanied by a despatch note are moved, it is the despatch note, or to be more exact the despatch note details, that appear on the data flow diagram. There is no mention of the goods.

It is also a common error to confuse data flows with flows of control. This is illustrated in Figure 2-15. Obviously what the designer of the diagram intended is that the exit data flow travels one way if there is sufficient stock, and the other way if there is not sufficient. It is not usual to indicate the conditions under which data flows on a data flow diagram. This would tempt the analyst to think of the system from the point of view of control rather than data flows.

The difference between a data store and a data flow often confuses the beginner. It is helpful to think of an analogy with water. Water flowing down a pipe is analogous to a data flow whereas water in a reservoir is the analogy for a data store.

■ DATA FLOW DIAGRAMS: AN EXAMPLE FROM KISMET

In systems investigation the analyst will have collected a great deal of information on tasks performed on document flows within the system. These tasks will be involved with the processing of business transactions and the generation of management information. The document processing will probably have been charted in a manual systems flowchart. In drawing data flow diagrams a logical approach is taken. It is important to ignore the location of processes in the manual system, the documents involved and who is responsible for the tasks. The designer should concentrate on the functions that are carried out.

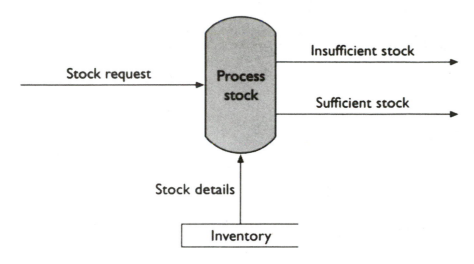

FIGURE 2-15   *Part of an Erroneous Data Flow Diagram*

If Kismet is considered, it can be seen that one very general course of action is taken. Orders from customers are processed to generate despatch notes (which eventually accompany goods sent to the customer) and to make up invoices (sent to the customer and to the accounts department). In doing this a price catalogue and customer account details are consulted and inventory records are updated.

This is indicated in Figure 2-16. In a more comprehensive analysis of all Kismet's functions there would be other interfacing subsystems such as purchasing, payroll, and accounting. If the analyst had seriously misunderstood the structure of the system this would be obvious at a glance.

Further progression can be made by breaking this order processing function into its component parts. There are three different types of tasks that occur when Kismet Ltd processes

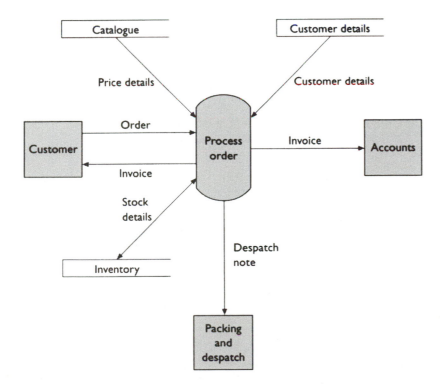

FIGURE 2-16   *A High-level View of Order Processing for Kismet*

a customer order. First, the company order is generated and approved. Second, stock is selected, inventory updated and a despatch note prepared. Finally, invoices are made up and sent to the customer and to accounts. Involved in this are various data flows, stores, processes, sources and sinks.

Read the case study and note where the sources, sinks, processes, stores and flows occur. These are:

1. **Sources/sinks:**
   (a) customer
   (b) credit control manager
   (c) packing and despatch department
   (d) accounting.

2. **Processes:**
   (a) generate approved company order
   (b) process stock transaction
   (c) make up invoice.

3. **Data stores:**
   (a) inventory
   (b) catalog
   (c) customer details
   (d) company order store
   (e) company order/invoice/despatch note.

4. **Data flows:**
   (a) customer order
   (b) company order
   (c) price details
   (d) stock details
   (e) customer credit details
   (f) invoice
   (g) customer invoice details
   (h) despatch note.

The data flow diagram at the first level can now be drawn (Figure 2-17). The following points should be noted:

1. The diagram can be thought of as starting at the top left and moving downwards and to the right. This gives some idea of the sequence of tasks performed, though it is not a rigid rule.

2. Each data flow, source/sink, process and store is labeled.

3. Certain tasks are left out of a data flow diagram. Any error-handling routine is usually omitted. For instance, although not stated, there would be a procedure for handling an incorrectly prepared company order form when the **customer name** and **customer#** were found not to correspond.

4. Departments and physical locations are ignored in the data flow diagram except where they appear as sources or sinks. For instance, although the processing of the stock transaction occurs in the stores and the generation of the approved company order occurs in the order department this information does not appear in the diagram.

5. Although the goods would accompany the despatch note to PACKING AND DESPATCH this is not shown on the data flow diagram as it is not a data flow.

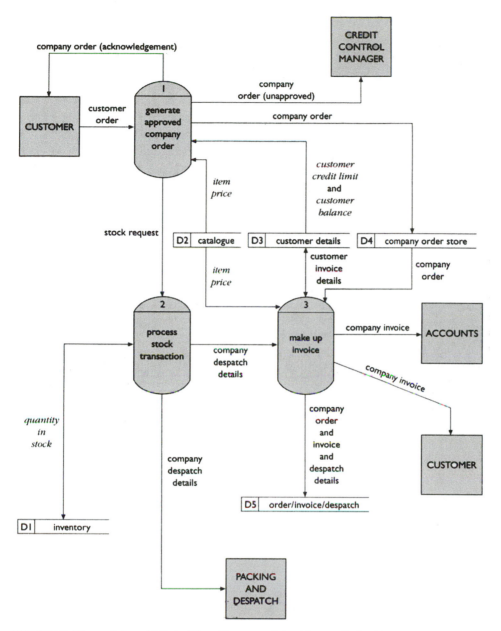

FIGURE 2-17   *A Level I Data Flow Diagram of the Kismet Order Processing System*

The data flow diagram shows a great deal about the flows of data between processes. However, it is important that the contents of data flows and the data stores are precisely specified as well. This will be of use when designing files or a database and when writing programs. A tabular representation of the contents of the various data elements appearing in the data flow diagram is shown in Figure 2-18.

The following points arising from this diagram should be noted:

1. The description in the case study does not go into much detail on the contents of each document or file (except in the case of the company order form). These contents are either implied by the nature of the task (**customer address** must be present to despatch the invoice) or should be found on the 'document description form' prepared by the analyst during investigation.

Source/sink	Process	
Customer	(1) Generate approved company order	
Credit control manager	(2) Process stock transaction	
Packing and despatch	(3) Make up invoice	
Accounting		

Data flow

Customer order	Company order	Company despatch details
customer #	*order#*	*despatch #*
customer name	*order date*	*order #*
[item #	*customer#*	*customer #*
item quantity]*	*customer name*	*customer name*
delivery address	*customer address*	*delivery address*
	delivery address	*despatch date*
Company invoice	*[item #*	*[item #*
invoice #	*item quantity*	*item quantity] **
invoice date	*item price]**	
customer #	*total*	
customer name		
customer address	Stock request	
order #	*order#*	
[item #	*customer#*	
item price	*customer name*	
item quantity]*	*delivery address*	
subtotal	*[item#*	
sales tax	*item quantity]*	
discount %		Customer invoice details
total payable		*customer#*
		customer name
		customer address
		turnover year to date

Data store

DI Inventory	D2 Catalog	D3 Customer details
item#	*item #*	*customer#*
quantity in stock	*item price*	*customer name*
.	.	*customer address*
.	.	*[delivery address]*
.	.	*customer balance*
.	.	*customer credit limit*
.	.	*turnover year to date*
.	.	*registration date*
D4 Company order store	D5 Order/invoice/despatch	.
see Company order	see Company order	.
	and invoice	.
	and despatch details	.

FIGURE 2-18 *Table of Descriptions of the Data Flow Diagram Elements for Kismet*

2. Extra detail will be stored for other tasks that are not part of the case as documented. For instance, under inventory there would be reorder levels. This omission does not matter as it would be remedied when the procedures for purchase and reorder of goods were incorporated in a data flow diagram.

3. The exact content of the data stores and flows will be recorded in a **data dictionary.** This is often described as a store of data about data. The dictionary is of considerable importance in analysis and design.

4. The meaning of [. . . .]* is that the contents of the brackets may be repeated an indeterminate number of times.

Data flow diagrams for simple systems are relatively straightforward to design. However for larger systems it is useful to follow a set of guidelines. These are:

1. Identify the major processes.
2. Identify the major data sources, sinks and stores.
3. Identify the major data flows.
4. Name the data flows, processes, sources, sinks and stores.
5. Draw the diagram.
6. Review the diagram, particularly checking that similar data flows, stores, sources and sinks have the same name and that different data flows and so on have different names.

■ DATA FLOW DIAGRAMS AT VARIOUS LEVELS

Two interconnected questions arise concerning data flow diagrams. These are:

1. What level of discrimination of processes should be shown on the data flow diagram? For instance, in Figure 2-17 the data process *generate approved company order* could be regarded as consisting of a number of subprocesses such as *accept order, check credit limit,* and *price order.* Should these be shown as processes?

2. What is the maximum number of processes that should be shown on a data flow diagram?

A major objective of a data flow diagram is its use in communication of the logical process model of the organization. It is difficult to understand a data flow diagram when it has more than seven to nine processes. This is the practical upper limit. If there is a tendency to overstep this then the data flow diagram should be redrawn, with processes that are logically grouped together being replaced by a single process that encompasses them all. The processes are not 'lost' from the model. They should appear on another data flow diagram that shows how this combined process can be exploded into its constituents. These constituents themselves may be complex and can be broken down into linked data processes shown on a data flow diagram at a lower level. This should be repeated until the processes are logically simple: that is until they cannot be broken down any further. This is illustrated in Figure 2-19.

The generation of levels in data flow diagrams has two other advantages. First, it naturally falls into line with the analyst's approach to top-down decomposition. That is, the analyst considers the major functions and processes first. These are then broken down into their constituents. The analyst can therefore concentrate on the higher-level data flow diagrams before designing the others. Second, the various levels correspond to the various degrees of detail by which the system is represented. This is useful for the analyst when discussing the results of the analysis with members of the organization. Senior management are more likely to be interested in a global view as given in a high-level data flow diagram. Other personnel will be concerned with more localized areas but in greater detail. For example, the person responsible for supervising the generation of approved company orders in Kismet will be interested in a lower-level, more detailed explosion of the *generate approved company order* process.

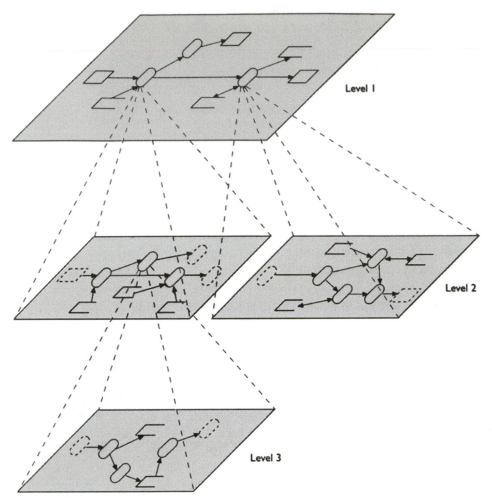

FIGURE 2-19   *Data Flow Diagrams at Various Levels*

The processes in the data flow diagram for Kismet (Figure 2-17) can be further decomposed. The generation of an approved company order is really a number of tasks:

1. Accept the order.
2. Prepare the company order form.
3. Price the goods.
4. Provisionally calculate the value of the order.
5. Check the credit-worthiness of the customer.

This is shown on a level 2 data flow diagram (Figure 2-20). The number 1 task in level 1 is exploded into seven tasks, 1.1–1.7. The inputs and outputs of this exploded chart must match those of the *parent* for which it is the functional decomposition.

There are two exceptions to this. First, certain local files need not be shown on the parent. In Figure 2-20 the *unapproved orders* store is such a file. It is only used as a temporary repository for the unapproved orders and does not feature in the rest of the system. Second, for the sake of clarity it may be necessary to amalgamate a number of data flows at the parent level. If functional analysis were carried out on function number 2, **process stock transaction,** the input/output data flow would be composed of an input flow with **item#** and **quantity in stock** as the data elements and an output with **item#, transaction type** and **item quantity** as elements.

If necessary, further analysis of the level 2 diagram could be undertaken by exploding chosen processes to a level 3 diagram.

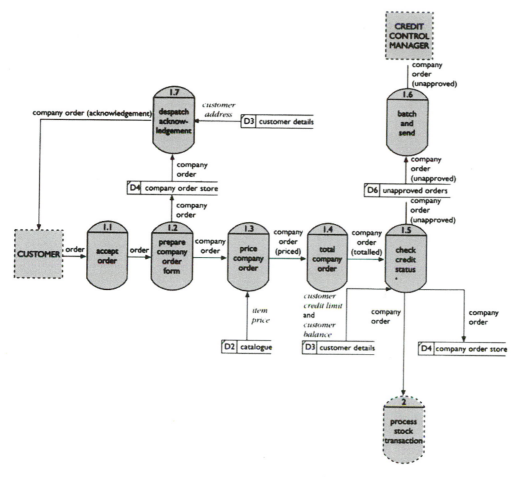

FIGURE 2-20   *A Level 2 Data Flow Diagram of the Kismet* Generate Approved Company
Order Process

▪ DESIGN COMMITMENTS OF THE DATA FLOW DIAGRAM APPROACH

It is important to separate the process of analysis from that of design. This may be difficult.
The process of analysis goes beyond description as every case of model building is partly a
process of design. Another aspect of design is that the choice of data flows, stores, processes,
sources and sinks is just one way of characterizing what is important in an organization's information system. To use these to build a model already commits the analyst to a certain design strategy: that of top-down, reductionist design.

However, within the design implications inherited through the use of structured techniques such as data flow diagrams, there is some separation of analysis from design. In the
case of Kismet the generation of an approved company order follows a certain (unnecessar-
ily repetitive) procedure, which is illustrated in the data flow diagram (Figure 2-20). In an
ideal design it is unlikely that the company order emanating from task 1.2 would be stored
in the company order file waiting for the trigger from the successful order-vetting (task 1.5)
before despatch of the acknowledgment. Rather, a copy of the approved company order form
would be sent to the customer as part of the output of task 1.5. In analysis the analyst considers what, logically, *is* done in order to carry out the functions of the organization. In
design the analyst states what, logically, *should* be done for the organization to carry out its
functions efficiently and effectively. The data flow diagram, by stripping away the physical
aspects of a system and revealing its logic at the appropriate level of detail, makes it easy to
see improvements that can be made.

The data flow design is an important tool of structured analysis and design. It ensures that a top-down approach is taken. This promotes a logical, as opposed to physical, view of data stores, flows and processes. In this way no premature commitment to physical aspects of design is made. Eventually the data stores *may* form part of the file structure or database of the system and the data processes *may* become the programs or subroutines. The flows will then correspond to the data passed in and out of programs and to and from the file structure or database. This though can be decided at a later stage.

The levels of data flow diagram correspond to the levels of top-down decomposition and the various levels of detail that management and users will need in order to discuss sensibly the analyst's understanding of processes.

The diagram may also be part of the agreed document provided at the end of analysis. Its non-technical nature means that it can be easily understood by users and agreed upon as a correct analysis of the system. This makes it a powerful communication tool, bridging the gap between analysts and users.

Finally, it makes it easy to sketch out different design alternatives for a future computerized system.

DESIGN

You have now reached the point in the analysis phase where you are ready to transform all of the information you have gathered and structured into some concrete ideas about the nature of the design for the new or replacement information system. This is called the *design strategy*. From requirements determination, you know what the current system does and you know what the users would like the replacement system to do. From requirements structuring, you know what forms the replacement system's process flow, process logic, and data should take, at a logical level independent of any physical implementation. For example, every data flow from a system process to the external environment represents an on-line display, printed report, or business form or document that the system can produce, often for use by a human. Every data flow from a data store to a process implies a retrieval capability of the system's files and database. Most processes represent a capability of the system to transform inputs into outputs. And data flows from the environment to system processes indicate capabilities to capture data on on-line displays, or some batch method, to validate the data and protect access to the system from the environment and to route raw data to the appropriate processing points.

Thus, at this point in the systems development process you have a preliminary specification of what the new information system should do and you understand why a replacement system is necessary to fix problems in the current system and to respond to new needs and opportunities to use information. Actually, there still may be some uncertainty about the capabilities of a new system. This uncertainty is due to competing ideas from different users and stakeholders on what they would like the system to do and to existing alternatives for an implementation environment for the new system. To bring analysis to a conclusion, your job is to take these structured requirements and transform them into several competing design strategies, one of which will be pursued in the design phase of the life cycle.

Part of generating a design strategy is determining how you want to acquire the replacement system using a combination of sources inside and outside the organization. If you decide to proceed with development in-house, you will have to answer general questions about software, such as whether all of the software should be built in-house or whether some software components should be bought off-the-shelf or contracted to software development companies. You will have to answer general questions about hardware and system software, such as whether the new system will run on a mainframe platform, stand-alone personal computers, or on a client/server platform, and whether the system can run on existing hardware. It is also not too early to begin thinking about data conversion issues, which must be addressed as you move from your current system to the new one. You even have to start thinking about how much training will be required for users, and how easy or difficult the system will be to implement. You have to determine whether you can build and implement the system you desire given the funding and management support you can count on. And you have to address these concerns for each alternative you generate. These issues need to be addressed so that you can update the Baseline Project Plan with detailed activities and resource requirements for the next life cycle phase—logical design—and probably for the physical design phase as well. That is, in this step of the analysis phase you bring the current phase to a close, prepare a report and presentation to management concerning continuation of the project, and get ready to move the project into the design phases.

In this section, you will learn why you need to come up with alternative design strategies and about guidelines for generating alternatives. You will then learn about the different issues that must be addressed for each alternative. Once you have generated your alternatives, you will have to choose the best design strategy to pursue. We include a discussion of one technique that analysts and users often use to decide among system alternatives and to help them agree on the best approach for the new information system.

Throughout this section we emphasize the need for sound project management. Now that you have seen the various techniques and steps of the analysis phase, we outline what a typical analysis phase project schedule might look like and discuss the execution of the analysis phase and the transition from analysis to design.

SELECTING THE BEST ALTERNATIVE DESIGN STRATEGY

Selecting the best alternative system involves at least two basic steps: (1) generating a comprehensive set of alternative design strategies and (2) selecting the one that is most likely to result in the desired information system, given all of the organizational, economic, and technical constraints that limit what can be done. In a sense then, the most likely strategy is the best one. A system design strategy is an approach to developing the system. The strategy includes the system's functionality, hardware and system software platform, and method for acquisition. We use the term design strategy in this section rather than the term alternative system because, at the end of analysis, we are still quite a long way from specifying an actual system. This delay is purposeful since we do not want to invest in design efforts until there is agreement on which direction to take the project and the new system. The best we can do at this point is to outline rather broadly the approach we can take in moving from logical system specifications to a working physical system. The overall process of selecting the best system strategy and the deliverables from this step in the analysis process are discussed next.

■ THE PROCESS OF SELECTING THE BEST ALTERNATIVE DESIGN STRATEGY

There are three subphases to systems analysis: requirements determination, requirements structuring, and generating alternative system design strategies and selecting the best one. After the system requirements have been structured in terms of process flow, process logic (decision and temporal), and data, analysts again work with users to package the requirements into different system configurations. Shaping alternative system design strategies involves the following processes:

- Dividing requirements into different sets of capabilities, ranging from the bare minimum that users would accept (the required features) to the most elaborate and advanced system the company can afford to develop (which includes all the features desired across all users). Alternatively, different sets of capabilities may represent the position of different organizational units with conflicting notions about what the system should do.

- Enumerating different potential implementation environments (hardware, system software, and network platforms) that could be used to deliver the different sets of capabilities. (Choices on the implementation environment may place technical limitations on the subsequent design phase activities.)

- Proposing different ways to source or acquire the various sets of capabilities for the different implementation environments.

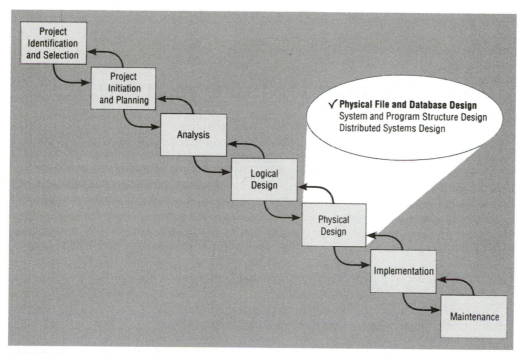

FIGURE 3-1 *Systems Development Life Cycle with the Physical Design Phase Highlighted*

In theory, if there are three sets of requirements, two implementation environments, and four sources of application software, there would be twenty-four possible design strategies. In practice, some combinations are usually infeasible or uninteresting. Further, usually only a small number—typically three—can be easily considered. Selecting the best alternative is usually done with the help of a quantitative procedure. Analysts will recommend what they believe to be the best alternative but management (a combination of the steering committee and those who will fund the rest of the project) will make the ultimate decision about which system design strategy to follow. At this point in the life cycle, it is also certainly possible for management to end a project before the more expensive phases of design and implementation are begun if the costs or risks seem to outweigh the benefits, if the needs of the organization have changed since the project began, or if other competing projects appear to be of greater worth and development resources are limited.

■ DELIVERABLES AND OUTCOMES

The primary deliverables from generating alternative design strategies and selecting the best one are outlined in Table 3.1. The primary deliverable that is carried forward into design is an updated Baseline Project Plan detailing the work necessary to turn the selected design strategy into the desired replacement information system. Of course, that plan cannot be assembled until a strategy has been selected, and no strategy can be selected until alternative strategies have been generated and compared. Therefore, all three objects—the alternatives, the selected alternative, and the plan—are listed as deliverables in Table 3.1. Further, these three deliverables plus

TABLE 3.1
Deliverables for Generating Alternatives and Selecting the Best One

1. At least three substantively different system design strategies for building the replacement information system

2. A design strategy judged most likely to lead to the most desirable information system

3. A Baseline Project Plan for turning the most likely design strategy into a working information system

the supporting deliverables from requirements determination and structuring steps are necessary information to conduct systems design, so all of this information is carried in the project dictionary and CASE repository for reference in subsequent phases.

GENERATING ALTERNATIVE DESIGN STRATEGIES

In many cases, it may seem to an analyst that the solution to an organizational problem is obvious. Typically, the analyst is very familiar with the problem, having conducted an extensive analysis of it and how it has been solved in the past, or the analyst is very familiar with a particular solution that he or she attempts to apply to all organizational problems encountered. For example, if an analyst is an expert at using advanced database technology to solve problems, then there is a tendency for the analyst to recommend advanced database technology as a solution to every possible problem. Or if the analyst designed a similar system for another customer or business unit, the "natural" design strategy would be the one used before. Given the role of experience in the solutions analysts suggest, analysis teams typically generate at least two alternative solutions for every problem they work on.

A good number of alternatives for analysts to generate is three. Why three? Three alternatives can neatly represent both ends and the middle of a continuum or spectrum of potential solutions. One alternative represents the low end of the spectrum. Low-end solutions are the most conservative in terms of the effort, cost, and technology involved in developing a new system. In fact, low-end solutions may not involve computer technology at all, focusing instead on making paper flows more efficient or reducing the redundancies in current processes. A low-end strategy provides all the required functionality users demand with a system that is minimally different from the current system.

Another alternative represents the high end of the spectrum. High-end alternatives go beyond simply solving the problem in question and focus instead on systems that contain many extra features users may desire. Functionality, not cost, is the primary focus of high-end alternatives. A high-end alternative will provide all desired features using advanced technologies which often allow the system to expand to meet future requirements. Finally, the third alternative lies between the extremes of the low-end and high-end systems. Such alternatives combine the frugality of low-end alternatives with the focus on functionality of high-end alternatives. Midrange alternatives represent compromise solutions. There are certainly other possible solutions that exist outside of these three alternatives. Defining the bottom, middle, and top possibilities allows the analyst to draw bounds around what can be reasonably done.

How do you know where to draw bounds around the potential solution space? The analysis team has already gathered the information it needs to identify the solution space, but first that information must be systematically organized. There are two major considerations. The first is determining what the minimum requirements are for the new system. These are the mandatory features, any of which, if missing, make the design strategy useless. Mandatory features are those that everyone agrees are necessary to solve the problem or meet the opportunity. Which features are mandatory can be determined from a survey of users and other stakeholders who have been involved in requirements determination. You would conduct this survey near the end of the analysis phase after all requirements have been structured and analyzed. In this survey, stakeholders rate features discovered during requirements determination or categorize features on some desirable-mandatory scale, and an arbitrary breakpoint is used to divide mandatory from desired features. Some organizations will break the features into three categories: mandatory, essential, and desired. Whereas mandatory features screen out possible solutions, essential features are the important capabilities of a system which will serve as the primary basis for comparison of different design strategies. Desired features are those that users could live without but which are used to select between design strategies that are of almost equal value in terms of essential features. Features can take many different forms. Features might include

- Data kept in system files (for example, multiple customer addresses so that bills can be sent to addresses different from where we ship goods)

- System outputs (printed reports, on-line displays, transaction documents—for example, a paycheck or sales summary graph)
- Analyses to generate the information in system outputs (for example, a sales forecasting module or an installment billing routine)
- Expectations on accessibility, response time, or turnaround time for system functions (for example, on-line, real-time updating of inventory files)

The second consideration in drawing bounds around alternative design strategies is determining the constraints on system development. Constraints may exist on such factors as

- A date when the replacement system is needed
- Available financial and human resources
- Elements of the current system that cannot change
- Legal and contractual restrictions
- The importance or dynamics of the problem which may limit how the system can be acquired (for example, a strategically important system that uses highly proprietary data probably cannot be outsourced or purchased)

Remember, be impertinent and question whether stated constraints are firm; you may want to consider some design alternatives that violate constraints you consider to be flexible.

Both requirements and constraints must be identified and ranked in order of importance. The reason behind such a ranking should be clear. Whereas you can design a high-end alternative to fulfill every wish users have for a new system, you design low-end alternatives to fulfill only the most important wishes. The same is true of constraints. Low-end alternatives will meet every constraint; high-end alternatives will ignore all but the most daunting constraints.

ISSUES TO CONSIDER IN GENERATING ALTERNATIVES

The required functionality of the replacement system and the constraints that limit that functionality form the basis for the many issues that must be considered in putting together all of the pieces that comprise alternative design strategies. That is, most of the substantive debate about alternative design strategies hinges on the relative importance of system features. Issues of functionality lead, however, to other associated issues such as whether the system should be developed and run in-house, software and hardware selection, implementation, and organizational limitations such as available funding levels. This list is not comprehensive, but it does remind you that an information system is more than just software. Each issue must be considered when framing alternatives. We will discuss each consideration in turn, beginning with the outsourcing decision.

■ OUTSOURCING

If another organization develops or runs a computer application for your organization, that practice is called outsourcing. Outsourcing includes a spectrum of working arrangements (*Business Week,* June 19, 1989; Moad, 1993). At one extreme is having a firm develop and run your application on their computers—all you do is supply input and take output. A common example of such an arrangement is a company that runs payroll applications for clients so that clients don't have to develop an independent in-house payroll system. Instead they simply provide employee payroll information to the company and, for a fee, the company returns completed paychecks, payroll accounting reports, and tax and other statements for employees. For many organizations, payroll is a very cost-effective operation when outsourced in this way. In another example of outsourcing arrangements, you hire a company to run your applications at your site on your computers. In some cases, an organization employing such an arrangement will dissolve some or all of its information systems unit and fire all of its information systems employees. Most of the time, though, the company brought in to run the organization's computing will hire many of the information systems unit employees.

Why would an organization outsource its information systems operations? As we saw in the payroll example, outsourcing may be cost-effective. If a company specializes in running payroll for other companies, it can leverage the economies of scale it achieves from running one very stable computer application for many organizations into very low prices. But why would an organization dissolve its entire information processing unit and bring in an outside firm to manage its computer applications? One reason may be to overcome operating problems the organization faces in its information systems unit. For example, the City of Grand Rapids, Michigan, hired an outside firm to run its computing 20 years ago in order to overcome personnel problems made difficult by union contracts and civil service constraints. Another reason for total outsourcing is that an organization's management may feel its core mission does not involve managing an information systems unit and that it might achieve more effective computing by turning over all of its operations to a more experienced, computer-oriented company. Kodak decided in the late 1980s that it was not in the computer applications business and turned over management of its mainframes to IBM and management of its personal computers to Businessland (Applegate and Montealegre, 1991).

Outsourcing is an alternative analysts need to be aware of. When generating alternative system development strategies for a system, you as an analyst should consult organizations in your area that provide outsourcing services. It may well be that at least one such organization has already developed and is running an application very close to what your users are asking for. Perhaps outsourcing the replacement system should be one of your alternatives. Knowing what your system requirements are before you consider outsourcing means that you can carefully assess how well the suppliers of outsourcing services can respond to your needs. However, should you decide not to consider outsourcing, you need to consider whether some software components of your replacement system should be purchased and not built.

■ SOURCES OF SOFTWARE

We can group organizations that produce software into four major categories: hardware manufacturers, packaged software producers, custom software producers, enterprise-wide solutions, and in-house developers.

HARDWARE MANUFACTURERS   At first it may seem counter-intuitive that hardware manufacturers would develop information systems or software. Yet hardware manufacturers are among the largest producers of software; for example, IBM is a leader in software development (Table 3.2). However, IBM actually develops relatively little application software (roughly 15 percent of their software revenue is from application software). Rather, IBM's leadership comes from its operating systems and utilities (like sort routines or database man-

TABLE 3.2
The Top 10 U.S. Software Companies in the 1996 *Datamation* 100

Rank among Software Companies	Company	Revenues from Software, 1996
1	IBM	$12,911
2	Microsoft	9,435
3	Hitachi	5,487
4	Fujitsu	4,755
5	Computer Associates	3,157
6	NEC	2,310
7	Oracle	2,280
8	SAP	1,700
9	Novell	1,225
10	Digital	1,225

Notes: Revenue in millions. Market share is percent of *Datamation* 100 software revenues.
Adapted from *Datamation*, July, 1997.

agement systems) for the hardware it manufactures, as well as its middleware, the software that links one set of software services to another.

PACKAGED SOFTWARE PRODUCERS   The growth of the software industry has been phenomenal since its beginnings in the mid-1960s. Now, some of the largest computer companies in the world, as measured by the *Datamation* 100, are companies that produce software exclusively (see Table 3.2). Consulting firms, such as American Management Systems and Andersen Consulting, also rank in the top 40 packaged software producers.

Software companies develop what are sometimes called *prepackaged* or *off-the-shelf systems*. Microsoft's Project and Intuit's Quicken™, QuickPay™, and QuickBooks™ are popular examples of such software. The packaged software development industry serves many market segments. Their software offerings range from general, broad-based packages, such as general ledger, to very narrow, niche packages, such as software to help manage a day care center. Software companies develop software to run on many different computer platforms, from microcomputers to large mainframes. The companies range in size from just a few people to thousands of employees. Software companies consult with system users after the initial software design has been completed and an early version of the system has been built. The systems are then beta-tested in actual organizations to determine whether there are any problems with the software or if any improvements can be made. Until testing is completed, the system is not offered for sale to the public. Unfortunately, the software is sometimes put on the market before it is ready. Figure 3-2 describes the problems Ashton-Tate encountered when it released dBASE IV too soon.

Some off-the-shelf software systems cannot be modified to meet the specific, individual needs of a particular organization. Such application systems are sometimes called turnkey

---

In October 1988, Ashton-Tate Corporation released its long-awaited version of dBASE IV™, the successor to its popular dBASE III™ database management system. dBASE IV was four times larger than dBASE III, with roughly 500,000 lines of code. Getting the program to operate on a standard IBM-compatible personal computer required ingenuity on the part of programmers. They had to figure out how to chop the program into pieces that could be swapped in and out of computer memory, a difficult task. The program was so complex, it outpaced Ashton-Tate's testing procedures at the time. The result: the company shipped a product containing thousands of errors. Customers complained dBASE IV was too slow and too likely to crash. Further, it lacked features customers were looking forward to having.

Ashton-Tate had once been considered one of the Big Three microcomputer software companies, along with Lotus Development Corp. and Microsoft Corp. In 1985, Ashton-Tate could boast of having 62.5% of the database market for microcomputers. In 1988, before the release of dBASE IV, Ashton-Tate controlled 43% of that market, but market share slipped to 35% in 1989 and to 30% in 1990. The company suffered through three quarterly losses in 1990, ending the year with a slight profit in the fourth quarter.

It took two years for Ashton-Tate to fix the problems with dBASE IV. They finally released a new version in the fall of 1990, two years after the fiasco. By that time, however, it was too late. They had lost many of their customers to fast-growing companies like Oracle Corp. To add to their problems, in December 1990, Ashton-Tate lost their copyright protection for their dBASE products. Ashton-Tate had sued rival Fox Software Inc. for copyright infringement. The judge in the case ruled that Ashton-Tate's products were initially based on a database system in the public domain, therefore invalidating the copyright protection they had for the products. Without such protection, Ashton-Tate had no right to sue Fox.

Ashton-Tate rewon their copyright protection for dBASE in 1991. That same year, however, the company was purchased by another rival, Borland International Inc.

Sources: Zachary, G. P. "How Ashton-Tate Lost Its Leadership in PC Software Arena."
Wall Street Journal, April 11, 1990, A1–A2.

Zachary, G. P. and Bulkeley, W. M. "Ashton-Tate Loses Flagship Software's Copyright Shield." Wall Street Journal, December 14, 1990, B1, B4.

FIGURE 3-2   *Ashton-Tate's dBase IV disaster*

systems. The producer of a *turnkey system* will only make changes to the software when a substantial number of users ask for a specific change. Other off-the-shelf application software can be modified or extended, however, by the producer or by the user, to more closely fit the needs of the organization. Even though many organizations perform similar functions, no two organizations do the same thing in quite the same way. A turnkey system may be good enough for a certain level of performance but it will never perfectly match the way a given organization does business. A reasonable estimate is that off-the-shelf software can at best meet 70 percent of an organization's needs. Thus, even in the best case, 30 percent of the software system doesn't match the organization's specifications.

CUSTOM SOFTWARE PRODUCERS   If a company has a need for an information system but does not have the expertise or the personnel available to develop the system in-house and a suitable off-the-shelf system is not available, the company will likely consult a custom software company. Consulting firms, such as Price Waterhouse or EDS, will help a firm develop custom information systems for internal use. These firms employ people with expertise in the development of information systems. Their consultants may also have expertise in a given business area. For example, consultants who work with banks understand financial institutions as well as information systems. Consultants use many of the same methodologies, techniques, and tools that companies use to develop systems in-house. The 10 largest world-wide computer-services firms (based on revenues not only of custom software development but also outsourcing and other services) are listed in Table 3.3. Other large U.S. firms are ADP, TRW, First Data, and AT&T.

ENTERPRISE SOLUTIONS SOFTWARE   More and more organizations are choosing complete software solutions, sometimes called enterprise solutions or enterprise resource

TABLE 3.3
Top 10 Worldwide Services and Support Suppliers for 1996 ($ millions)

1996 Rank	Company	1996 Services & Support Revenue
1	IBM	$22,785
2	Electronic Data Systems	14,441
3	Hewlett-Packard	9,462
4	Digital Equipment	5,988
5	Computer Sciences	5,400
6	Andersen Consulting	4,878
7	Fujitsu	4,160
8	Cap Gemini Sogeti	4,104
9	Unisys	3,950
10	ADP	3,567

Notes: Revenues in millions. Adapted from *Datamation*, July 1, 1997

planning systems, to support their operations and business processes. These software solutions consist of a series of integrated modules. The modules pertain to specific, traditional business functions, such as accounting, distribution, manufacturing, and human resources. The difference between the modules and traditional approaches is that the modules are integrated to focus on business processes rather than on business functional areas. Using enterprise software solutions, a firm can integrate all parts of a business process, such as order entry and filling, in a unified information system. All aspects of a single transaction, including receiving the order, picking goods in a warehouse, shipping, inventory adjustments, billing, and accounts receivable, occur seamlessly within a single information system, rather than in a series of disjointed, separate systems focused on business functional areas.

The benefits of the enterprise solutions approach include a single repository of data for all aspects of a business process and the flexibility of the modules. A single repository implies more consistent and accurate data, as well as less maintenance. The modules are very flexible because additional modules can be added as needed, once the basic system is in place. Added modules are immediately integrated into the existing system.

There are also disadvantages to enterprise solutions software. The systems are very complex, so implementation can take a long time to complete. Because the systems are complex, organizations typically do not have the necessary expertise in-house to implement the systems. Instead, they must rely on consultants or employees of the software vendor, and such expert help can be very expensive. In some cases, organizations must change how they do business in order to benefit from a migration to enterprise solutions.

There are three major vendors of enterprise solution software. The best known is probably SAP AG, a German firm, known for its flagship product R/3. SAP stands for Systems, Applications, and Products in Data Processing. SAP AG was founded in 1972, but most of its growth has occurred since 1992. In 1996, SAP AG was the eighth largest supplier of software in the world (see Table 3.2). Another supplier of enterprise solution software is People-Soft, Inc., a U.S. firm founded in 1987. PeopleSoft began with enterprise solutions that focused on human resources management but has since expanded to cover financials, materials management, distribution, and manufacturing. The third vendor is The Baan Company, a Dutch firm founded in 1978. Baan's applications include manufacturing, finance, distribution and transportation, and service. The most recent offerings from all three vendors run on a client-server platform.

IN-HOUSE DEVELOPMENT We have talked about three different types of external organizations that serve as sources of software, but in-house development remains an option. Of course, in-house development need not entail development of all of the software that will comprise the total system. Hybrid solutions involving some purchased and in-house software components are common. Table 3.4 compares the different software sources.

If you choose to acquire software from outside sources, this choice is made at the end of the analysis phase. Choosing between a package or external supplier will be driven by your needs, not by what the external party has to sell. As we will discuss, the results of your analysis study will define the type of product you want to buy and will make working with an external supplier much easier, productive, and worthwhile.

TABLE 3.4
Comparison of Four Different Sources of Software Components

Producers	Source of Application Software?	When to Go to This Type Organization for Software	Internal Staffing Requirements
Hardware Manufacturers	Generally not	For system software and utilities	Varies
Packaged Software Producers	Yes	When supported task is generic	Some IS and user staff to define requirements and evaluate packages
Custom Software Producers	Yes	When task requires custom support and system can't be built internally	Internal staff may be needed, depending on application
In-House Developers	Yes	When resources and staff are available and system must be built from scratch	Internal staff necessary though staff size may vary

■ CHOOSING OFF-THE-SHELF SOFTWARE

Once you have decided to purchase off-the-shelf software rather than write some or all of the software for your new system, how do you decide what to buy? There are several criteria to consider, and special criteria may arise with each potential software purchase. For each criterion, an explicit comparison should be made between the software package and the process of developing the same application in-house. The most common criteria are as follows:

- Cost
- Functionality
- Vendor support
- Viability of vendor
- Flexibility
- Documentation
- Response time
- Ease of installation

These criteria are presented in no particular order. The relative importance of the criteria will vary from project to project and from organization to organization. If you had to choose two criteria that would always be among the most important, those two would probably be vendor viability and vendor support. You don't want to get involved with a vendor that might not be in business tomorrow. Similarly, you don't want to license software from a vendor with a reputation for poor support. How you rank the importance of the remaining criteria will very much depend on the specific situation in which you find yourself.

Cost involves comparing the cost of developing the same system in-house to the cost of purchasing or licensing the software package. You should include a comparison of the cost of purchasing vendor upgrades or annual license fees with the costs you would incur to maintain your own software. Costs for purchasing and developing in-house can be compared based on economic feasibility measures (for example, a present value can be calculated for the cash flow associated with each alternative). Functionality refers to the tasks the software can perform and the mandatory, essential, and desired system features. Can the software package perform all or just some of the tasks your users need? If some, can it perform the necessary core tasks? Note that meeting user requirements occurs at the end of the analysis phase because you cannot evaluate packaged software until user requirements have been gathered and structured. Purchasing application software is not a substitute for conducting the systems analysis phase; rather, purchasing software is part of one design strategy for acquiring the system identified during analysis.

As we said earlier, vendor support refers to whether and how much support the vendor can provide. Support occurs in the form of assistance to install the software, to train user and systems staff on the software, and to provide help as problems arise after installation. Recently, many software companies have significantly reduced the amount of free support they will provide customers, so the cost to use telephone, on-site, fax, or computer bulletin board support facilities should be considered. Related to support is the vendor's viability. You don't want to get stuck with software developed by a vendor that might go out of business soon. This latter point should not be minimized. The software industry is quite dynamic, and innovative application software is created by entrepreneurs working from home offices—the classic cottage industry. Such organizations, even with outstanding software, often do not have the resources or business management ability to stay in business very long. Further, competitive moves by major software firms can render the products of smaller firms outdated or incompatible with operating systems. One software firm we talked to while developing this book was struggling to survive just trying to make their software work on any supposedly IBM-compatible PC (given the infinite combination of video boards, monitors, BIOS chips, and other components). Keeping up with hardware and system software change may be more than a small firm can handle, and good off-the-shelf application software is lost.

Flexibility refers to how easy it is for you, or the vendor, to customize the software. If the software is not very flexible, your users may have to adapt the way they work to fit the software. Are they likely to adapt in this manner? Purchased software can be modified in several ways. Sometimes, the vendor will be willing to make custom changes for you, if you are willing to pay for the redesign and programming. Some vendors design the software for customization. For example, the software may include several different ways of processing data and, at installation time, the customer chooses which to initiate. Also, displays and reports may be easily redesigned if these modules are written in a fourth-generation language. Reports, forms, and displays may be easily customized using a process whereby your company name and chosen titles for reports, displays, forms, column headings, etc. are selected from a table of parameters you provide. You may want to employ some of these same customization techniques for in-house developed systems so that the software can be easily adapted for different business units, product lines, or departments.

Documentation includes the user's manual as well as technical documentation. How understandable and up-to-date is the documentation? What is the cost for multiple copies, if required? Response time refers to how long it takes the software package to respond to the user's requests in an interactive session. Another measure of time would be how long it takes the software to complete running a job. Finally, ease of installation is a measure of the difficulty of loading the software and making it operational.

VALIDATING PURCHASED SOFTWARE INFORMATION   One way to get all of the information you want about a software package is to collect it from the vendor. Some of this information may be contained in the software documentation and technical marketing literature. Other information can be provided upon request. For example, you can send prospective vendors a questionnaire asking specific questions about their packages. This may be part of a request for proposal (RFP) or request for quote (RFQ) process your organization requires when major purchases are made. Space does not permit us to discuss the topic of RFPs and RFQs here; you may wish to refer to purchasing and marketing texts if you are unfamiliar with such processes.

There is, of course, no replacement for actually using the software yourself and running it through a series of tests based on the criteria for selecting software. Remember to test not only the software but also the documentation, training materials, and even the technical support facilities. One requirement you can place on prospective software vendors as part of the bidding process is that they install (free or at an agreed-upon cost) their software for a limited amount of time on your computers. This way you can determine how their software works in your environment, not in some optimized environment they have.

One of the most reliable and insightful sources is other users of the software. Vendors will usually provide a list of customers (remember, they will naturally tell you about satisfied customers, so you may have to probe for a cross section of customers) and people who are willing to be contacted by prospective customers. And here is where your personal network of contacts, developed through professional groups, college friends, trade associations, or local business clubs, can be a resource; do not hesitate to find some contacts on your own. Such current or former customers can provide a depth of insight on use of a package at their organizations.

To gain a range of opinion about possible packages, you can use independent software testing and abstracting services that periodically evaluate software and collect user opinions. Such surveys are available for a fee either as subscription services or on demand (two popular services are Auerbach Publishers and DataPro); occasionally, unbiased surveys appear in trade publications. Often, however, articles in trade publications, even software reviews, are actually seeded by the software manufacturer and are not unbiased.

If you are comparing several software packages, you can assign scores for each package on each criterion and compare the scores using the quantitative method we demonstrate at the end of the section for comparing alternative system design strategies.

■ HARDWARE AND SYSTEMS SOFTWARE ISSUES

The first question you need to ask yourself about hardware and system software is whether the new system that follows a particular design strategy can be run on your firm's existing hardware and software platform. System software refers to such key components as operating systems, database management systems, programming languages, code generators, and network software. To determine if current hardware and system software is sufficient, you should consider such factors as the age and capacity of the current hardware and system software, the fit between the hardware and software and your new application's goals and proposed functionality and, if some of your system components are off-the-shelf software, whether the software can run on the existing hardware and system software. The advantages to running your new system on the existing platform are persuasive:

1. Lower costs as little, if any, new hardware and system software has to be purchased and installed.

2. Your information systems staff is quite familiar with the existing platform and how to operate and maintain it.

3. The odds of integrating your new application system with existing applications are enhanced.

4. No added costs of converting old systems to a new platform, if necessary, or of translating existing data between current technology and the new hardware and system software you have to acquire for your system.

On the other hand, there are also very persuasive reasons for acquiring new hardware or system software:

1. Some software components of your new system will only run on particular platforms with particular operating systems.

2. Developing your system for a new platform gives your organization the opportunity to upgrade or expand its current technology holdings.

3. New platform requirements may allow your organization to radically change its computing operations, as in moving from mainframe-centered processing to a database machine or a client-server architecture.

As the determination of whether or not to acquire new hardware and system software is so context-dependent, providing platform options as part of your design strategy alternatives is an essential practice.

At this point, if you decide that new hardware or system software is a strong possibility, you may want to issue a request for proposal (RFP) to vendors. The RFP will ask the vendors to propose hardware and system software that will meet the requirements of your new system. Issuing an RFP gives you the opportunity to have vendors carry out the research you need in order to decide among various options. You can request that each bid submitted by a vendor contain certain information essential for you to decide on what best fits your needs. For example, you can ask for performance information related to speed and number of operations per second. You can ask about machine reliability and service availability and whether there is an installation nearby which you can visit for more information. You can ask to take part in a demonstration of the hardware. And of course the bid will include information on cost. You can then use the information you have collected in generating your alternative design strategies.

■ IMPLEMENTATION ISSUES

Implementing a new information system is just as much an organizational change process as it is a technical process. Implementation involves more than installing a piece of software, turning it on, and moving on to the next software project. New systems often entail new ways of performing the same work, new working relationships, and new skills. Users have to be trained. Disruptions in work procedures have to be found and addressed. In addition, system implementation may be phased in over many weeks or even months. You must address the technical and social aspects of implementation as part of any alternative design strategy. Management and users will want to know how long the implementation will take, how much training will be required, and how disruptive the process will be.

■ ORGANIZATIONAL ISSUES

One reason management is so interested in the outlook for implementation is that all of the concerns we listed previously cost management money. The longer the implementation process, the more training required; the more disruption expected, the more it will cost to implement the system. Implementation costs are just one cost management has to consider for the new system. Management must also consider the costs of the design process that precedes implementation and the cost of maintaining the system once implementation is over. Overall cost and the availability of funding is just one of the organizational issues to consider in developing alternative design strategies.

A second organizational issue is determining what management will support. Even if adequate funding is available, your organization's management may not be willing or able to support one or another of your alternatives. For example, if your new system differs dramatically from what the corporate office has determined is adequate or from the corporate standard computing environment, your local management may not be willing to support a system that will irritate the corporate office. Most organizations like to have a manageably small number of basic technologies since only a few choices on hardware and software platforms can be supported well. If your new system calls for high levels of cooperation across departments when current operations involve very little of such cooperation, your management may not be willing to support such a system. Your management may also have politically inspired reasons not to support your new system. For example, if your information systems unit reports to the organization's chief financial officer and your new system strengthens a rival department such as manufacturing, management may not support such a system because they prefer to continue the status quo whereby finance is stronger than manufacturing.

A third organizational issue is the extent to which users will accept the new system and use it as designed. A high-end, high-technology solution that represents a radical break in what users are familiar with may have less of a chance of acceptance than a system that is closer to what users know and use. On the other hand, acceptance of a high-end alternative will depend on the users. Some users may demand nothing less than what leading-edge technology can deliver.

DETAILED DESIGN

Data flow diagrams were used to develop a high-level logical model of the processes and the data flows between them in the system. They were particularly helpful in sketching design alternatives using automation boundaries. The logic of the processes themselves has been captured in structured tools such as decision tables, structured English, or logic flowcharts. The data model of the organization has been produced using entity-relationship modeling.

The use of this logical approach has ensured that no premature physical design decisions have been taken. The concentration has always been on the question 'What logically is required of the system and what logically must be achieved in order to do this?' not 'How are we physically going to accomplish this?' However, the time has come to move away from these logical models towards the detailed physical design of the computerized system (see Figure 3-3). The data flow diagrams and logic representations developed so far will be invaluable in producing program specifications. The entity model will be essential in designing a database model.

Any system can be considered as being composed of input, output, storage and processing elements. This approach is used in looking at the various tasks to be covered in detailed systems design. Control features figure prominently in the design of all the other elements.

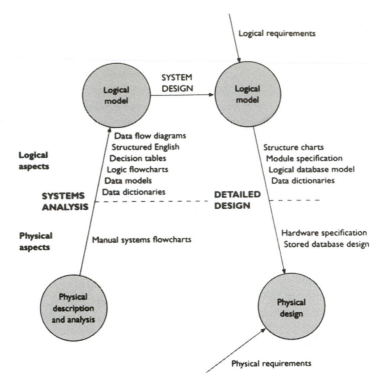

FIGURE 3-3 *Tools Used During the Stages of Systems Analysis and Design*

■ PROCESS DESIGN

Process design covers the need to design and specify the processing hardware, and to design and specify the software for running in the central processing unit.

The specification of processor hardware is a technical skill beyond the scope of this text. The analyst will need to consider the demands to be made on the processor. In particular, the volumes of transactions to be processed per day and the processing requirement for each, the speed of response needed for interactive enquiries, the number of simultaneous users and the types of peripheral devices required, the amount of RAM, the complexity of programs and the extent to which the system should be able to cope with future increases in demand are all determinants of the processing power required. Decisions taken on whether to centralize or distribute computing and the mix between batch and online processing will all affect the decision.

■ DEVELOPMENT OF THE STRUCTURE CHART AND MODULE SPECIFICATION FOR KISMET

The level 1 data flow diagram for the Kismet order processing system is given again in Figure 3-4(a) and the level 2 breakdown in Figure 3-5(a). These will be used to generate a (partial) structure chart for the Kismet system.

The main part of the level 1 data flow diagram suggests that there are three separate functions. The first deals with the generation of approved orders. The second processes stock transactions to update stock records and produce despatch notes. Finally, invoices are made up and despatched. This can be seen in the simple structure chart of Figure 3-4(b).

The structure chart dealing with the generation of the approved orders is shown in Figure 3-5(b). As is common in the production of hierarchical structure charts the data flows between the various modules are omitted. This is usually done to simplify the chart. There is, however, another reason. The structure is prepared before the individual module specification and it is at that point that the analyst's attention is devoted to the precise description of the data flows. Structure charts are then first derived without all the data flows.

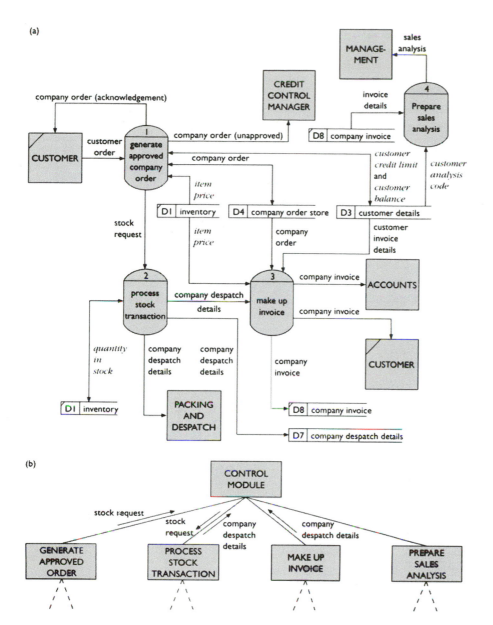

FIGURE 3-4   (a) A Data Flow Diagram of Kismet's Order Processing System
(b) Part of a Structure Chart for the Kismet Order Processing System

Consideration of the data flow diagram in Figure 3-5(a) indicates that the process of generating an approved order includes three important parts: preparing the priced company order, checking the credit status of the customer for the order and, depending on the result of the credit check, handling the order in one way or another. This is shown by the second-level decomposition of the tasks in the structure chart (Figure 3-5(b)).

Preparing the priced order consists of two tasks. First, the order details are obtained through keyboard entry. This is shown by the module GET ORDER. Second, the order details are then used to value the order (VALUE ORDER). This itself consists of two tasks. The first is to obtain the price details for each item ordered and the second is to calculate the total value of the order. Note that preparing the order involves the two processes GET ORDER and VALUE ORDER in that sequence. This is just the sequence of the two corresponding data processes in the data flow diagram—process 1.1 and process 1.2 (Figure 3-5(a)).

The next process in the chain, process 1.3, is performed by the next module to be executed—CHECK CREDIT STATUS.

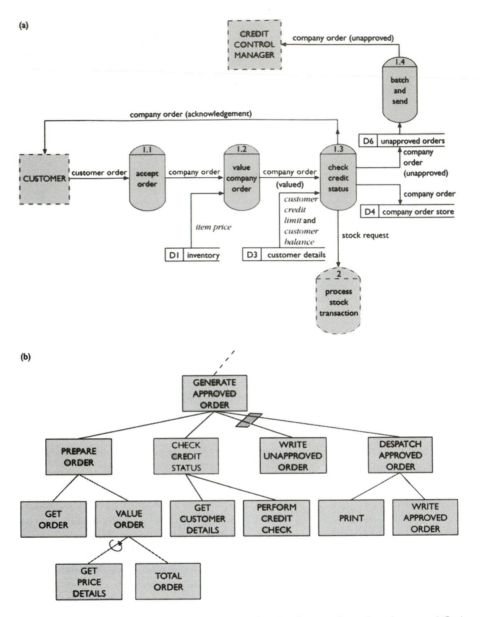

FIGURE 3-5   (a) The Data Flow Diagram for the Generation of an Approved Order
(b) A Structure Chart for the Generation of an Approved Order

Finally, depending on the outcome of the credit check, the order is stored (unapproved) in the store D.6 or printed for despatch to the customer and stored (approved) in the store D.4. This corresponds to the alternatives shown on the structure chart (Figure 3-5(b)).

The approved order is also needed by the module concerned with processing the stock transaction. It is passed, minus its price details, as the output 'stock request' of the GENERATE APPROVED ORDER module via the control module.

A sample of a module specification is given in Figure 3-6. The data inputs and outputs will be defined in the data dictionary. Note that the modules called and the module that calls it determine the position of the module CHECK CREDIT STATUS in the hierarchy chart.

SYSTEM	ORDER PROCESSING
MODULE NAME	CHECK CREDIT STATUS
AUTHOR	J. SMITH
DATE	3/3/98
MODULE CALLS	GET CUSTOMER DETAILS PERFORM CREDIT CHECK
MODULE IS CALLED BY	GENERATE APPROVED ORDER
MODULE INPUTS	*order total* *customer=*
MODULE OUTPUTS	*company order approval flag*
PROCESS	DO GET CUSTOMER DETAILS DO PERFORM CREDIT CHECK

FIGURE 3-6   *A Module Specification for CHECK CREDIT DETAIL*

The way a data flow diagram is used for the preparation of a structure chart can be summarized as follows:

1. View the diagram to establish whether the data processes fall into groups each associated with a type of process. View the resulting grouped collections of data processes to establish the key functions in each group. If each group has only a single data process within it this is straightforward.

2. Break down these functions to obtain modules that are highly cohesive and loosely coupled. Note that the sequence of data processes will indicate the left-to-right ordering of modules. Data flows that are conditional on a test performed in a data process will correspond to alternative module selections. This can be indicated using the diamond symbol. Lower-level data flow diagrams are to be used in this decomposition.

3. Now specify each module separately ensuring that named data items and data collections are entered in the data dictionary and that the data produced by major modules corresponds to the data being passed between processes in the data flow diagrams.

This gives a general idea of the development of a hierarchical structure chart and module specification. Do not form the impression that once the data flow diagrams have been developed the production of the modular structure chart is a mechanical task. It is not. The analysis of the high-level diagrams can often be difficult. Most important is the realization that to discover key functions and to decompose these into their constituents, each of which performs a single logical function that is translatable into a cohesive module, requires more than an uncritical reading of a data flow diagram. It is necessary for the analyst to obtain a deep appreciation of the system and the functions it performs. Although structured methods attempt to replace the art of design by the science of engineering there is still a place for creative understanding.

■ DATA STORE DESIGN

During systems analysis and design a conceptual data model of the organization was developed. This was based on modelling entities and their relationships. Later, attributes of the entities were established. A normalized entity model was then developed to a third normal form. The resulting normalized entities have been checked by deriving functional entity models to ensure that the various processes can be served with the data that they need. During systems design a decision will have been made as to whether to service the future data needs of the organization via a database approach using a database management system or via an application-led file-based approach. Either way, detailed design of the data store will be required and this will be aided by the normalized data model.

FILE-BASED SYSTEMS   In general, a file-based approach will have been chosen if there is little or no common data usage between applications and there is a need for fast transaction processing. The analyst will need to design record layouts. This involves specifying the names of the fields in each record, their type (for example, text or numeric) and their length. Records may be fixed or variable length. In the latter case this is either because fields are allowed to vary in length or because some fields are repeated. Record structures can become quite complicated if groups of fields are repeated within already repeating groups of fields.

As well as record design the storage organization and access for each file must be decided. This will determine the speed of access to individual records and the ease with which records can be added and deleted within the file. The most common types of storage organization are sequential, indexed, indexed-sequential, random storage with hashing, lists and inverted files. Different file organizations have different characteristics with respect to ease and speed of data storage and retrieval. Depending on the type of application the most appropriate organization is chosen.

The analyst will need to specify suitable backing-store hardware. This may involve tapes and tape read/write devices but nowadays is more likely to be disk-based. In order to select appropriate hardware the analyst must take into account the number of files and within each file the number of records and the storage required for each, the file organization chosen, the response time required, and the likely future developments, especially in terms of increases in size. These characteristics will allow the analyst to calculate the total storage requirements and to specify appropriate hardware.

SYSTEMS SPECIFICATION

Detailed design results in a systems specification. This is a comprehensive document describing the system as it is to be produced. There is no agreed format for this specification but it is likely to include the following:

- An executive summary: this provides a quick summary of the major points in the specification.
- A description of the proposed system and especially its objectives. Flow block diagrams and data flow diagrams can be used. The work to be carried out by the system and the various user functions should be covered.
- A complete specification of:
  - Programs: These will include module specifications and structure charts, together with test data.
  - Input: this will include specimen source documents, screen layouts, menu structures, control procedures.
  - Output: this will include specimen output reports, contents of listings, and so on.
  - Data storage: this is the specification of file and database structure.
- A detailed specification of controls operating over procedures within the system.

- A specification of all hardware requirements and performance characteristics to be satisfied.
- A specification of clerical procedures and responsibilities surrounding the system.
- A detailed schedule for the implementation of the system.
- Cost estimates and constraints.

The systems specification fulfils a number of roles. First, it is used as the exit criterion from the stage of detailed design prior to the stage of implementation. The systems specification is agreed by senior management, often the steering committee. Once accepted, large amounts of money are allocated to the project. This is necessary to purchase hardware, to code software, and to carry out physical installation. Second, the specification acts as source documentation from which programs are written and hardware tenders are arranged. Third, the document acts as a historical record of the system for future users and developers. Finally, it is used in the assessment of the system once the system is being used. Does the system meet its response times? Is it built according to design? These are examples of questions that can only be answered in consultation with the systems specification.

CASE AND OBJECT-ORIENTED APPROACHES TO ANALYSIS AND DESIGN

Systems analysis and design is not a static discipline. New approaches and methodologies are being produced regularly. Many of these are variations on old themes—the themes covered in the earlier parts of this section. Some, though, are different in nature. This section examines, amongst others, computer-aided software engineering and object-oriented analysis and design. Computer-aided software engineering (CASE) provides new tools for the speedy development of reliable computer systems. It is not uncommon for these to be based around object-oriented approaches to analysis and design. Object-oriented methods are having a large impact because they mesh with developments in other areas—particularly with object-oriented programming languages, object-oriented databases and, not least, CASE. Although prototyping has been covered earlier in connection with end-user computing and decision support systems its influence can be felt in other areas. In particular it may be used with the design of systems using CASE. While object-oriented methods currently have little in common with end-user computing there is a thread running through CASE and prototyping, which is likely to be emphasized in the coming years.

■ CASE (COMPUTER-AIDED SOFTWARE ENGINEERING)

The traditional cost curve for the design and development of a business information system locates most of the cost as falling within the area of implementation, particularly coding. Fourth-generation languages (4GLs) and other applications generation tools are an attempt to cut the cost in this area by enabling speedy development of systems through the automation of the production code.

The success of this can be measured by the fact that it is now possible to build a system and, if it fails to meet user needs, to redevelop it quickly and cheaply. This, in essence, is the philosophy behind prototyping. A problem with the approach is that it diminishes the role of formal requirements specification and systems design. This may be acceptable or even desirable for small systems, particularly for decision support, but the approach cannot cope with larger projects unless considerable design effort has been undertaken.

Structured methodologies, developed in the late 1970s and early 1980s, and covered elsewhere in this book, were a significant improvement in approaches to systems analysis and design as compared with traditional methods. By clearly identifying stages, tools and techniques, and documentation standards, consistency and quality control were maintained for systems design projects. Design errors were significantly reduced.

However, as emphasized earlier in this section, requirements analysis, especially if linked to strategic requirements, is not readily accessible to structured techniques. Errors still occur as a result of inaccurate requirements analysis and specifications (see Figure 3-7). Changes in requirements mean lengthy redesign. But just as the computer itself has been utilized to assist the coding process (4GLs and application generators) the next step was to use the computer in the requirements specification, analysis and design process itself. This is the place of computer-aided software (systems) engineering (CASE) (Figure 3-8).

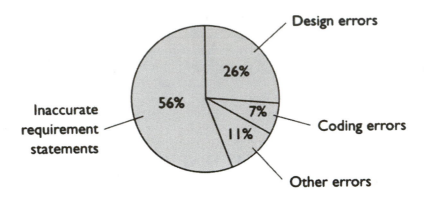

FIGURE 3-7   *Sources of Error in Implemented Applications (Parkinson, 1989)*

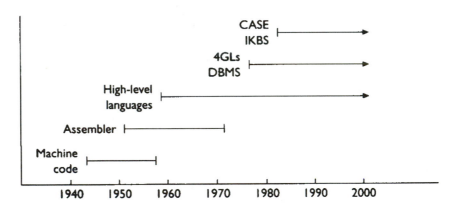

FIGURE 3-8   *The Place of CASE Tools in Software Development*

CASE automates various stages of the analysis and design process with the assistance of software tools. The CASE philosophy involves the use of these software tools working together with a methodology of systems analysis and design to:

- develop models that describe the business;
- aid in corporate planning;
- provide support for systems specification, documentation and design;
- aid in the implementation process.

CASE tools are used by computer professionals for carrying out some part of the systems analysis/design process. They are not end-user orientated though may be used in conjunction with end-users. The tools are normally run on high-performance development workstation PCs.

■ CASE SUPPORT

CASE tools provide assistance in the following ways (see Figure 3-9):

1. *Corporate planning of information systems:* Software is used to describe the organization, its goals, resources, responsibilities and structure. These descriptions may be used to create strategic plans. The software will be used to indicate and track relationships between the various components of the plan. The planning specifications will identify activities for incorporation in information systems developments.

2. *Creating specifications of requirements:* This corresponds to the non-strategic stages of systems analysis. At this stage an information system is analyzed into its component activities and data requirements. Many diagrammatic tools familiar to the reader from the stages of structured process and data analysis are used—for example, data flow diagrams and entity models.

3. *Creating design specifications:* CASE tools can be used to specify a design for a new system. These tools include software for producing HIPO charts, structured module specifications, and decision tables.

4. *Code-generation tools:* CASE support in this area will accept the outputs of the design specification and produce computer-generated code for direct execution.

5. *An information repository:* Central to any CASE assistance in systems analysis, design and implementation is the information repository. This stores information on entities, processes, data structures, business rules, source code and project management data. It is important, as at any one stage of development consistency and completeness must be maintained in descriptions of data, processes, rules and so on. Throughout the various stages of analysis and design it is vital that as each component develops from, for example, an item in a high-level plan through to a record type in a database, it can be tracked. In short the information repository does for the CASE process what the data dictionary, in a more limited way, does for structured process analysis and design.

6. *A development methodology:* Underlying any CASE approach is a development methodology. Indeed the first CASE tools were designed to provide automated diagramming facilities for data flow diagrams and entity relationship models within a structured methodology.

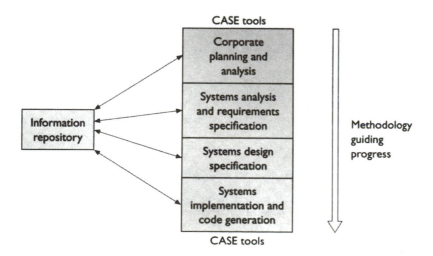

FIGURE 3-9   *CASE Assistance in the Systems Development Process*

■ CASE TERMINOLOGY

There are no agreed international or *de facto* standards on CASE. This is a problem not least because terminology can be used in widely varying ways. However, the following are in common use:

FRONT-END CASE/UPPER CASE/ANALYST WORKBENCH This refers to a set of tools that aid productivity in the analysis, design and documentation of systems. These tools are used to define the systems requirements and the systems properties. Typically the outputs of front-end CASE are:

- process and data structure specifications: data flow diagrams, state transition diagrams, flowcharts, entity-relationship diagrams, data dictionaries, pseudocode, structured English, decision tables, decision trees, structure charts, module specifications, Warnier-Orr diagrams
- screen definitions
- report definitions.

In order to achieve this there will be a wide variety of highly interactive screen-based prototyping facilities including:

- diagram editors
- screen painters
- dialogue prototyping aids.

BACK-END CASE/LOWER CASE/CODE GENERATOR/4GL GENERATOR This refers to tools that automate the latter stages of the implementation process. As input, back-end CASE tools may take the output specifications of front-end CASE tools. The outputs of back-end CASE are:

- source and object program code
- database and file definitions
- job control language.

LIFE CYCLE CASE/I-CASE This encapsulates tools for covering the entire process of corporate analysis and planning, systems analysis and design, and implementation. For example the information engineering methodology of James Martin is incorporated in several I-CASE tools.

REVERSE ENGINEERING/RE-ENGINEERING Reverse engineering is the opposite of the standard implementation process for the production of code. Reverse engineering takes existing unstructured code as input (often COBOL programs) and produces restructured code as output. In particular the reverse engineering process produces output where:

- Common subroutines and modules are identified.
- Conditional branches and loops are simplified in structure.
- Blocks of code are collapsed into single-line statements for ease of reference and use.
- Subroutines are called in hierarchies.
- New source code is generated fitting the above conditions.

The need for reverse engineering is generated by the recognition that most software used within an organization needs enhancement from time to time. Unless this was produced using a rigorous methodology (and many systems still in use based on COBOL code were not) the task facing the programmer is immense. Complex spaghetti logic structures are impossible or, at best, time-consuming to disentangle and alter. If the software has undergone amendment the side-effects of code alteration are not easily identifiable. The software tends to be bug-ridden and difficult to validate. Reverse engineering creates code that is structured and easily amendable. It thus enables existing software to be enhanced rather than abandoned when update is necessary.

■ CASE BENEFITS

CASE has grown in use over the last decade and is predicted to continue to increase in importance over the next decade. Some of the benefits of CASE are as follows:

1. *Enhancement of existing applications:* This can occur in two ways. First, systems produced using CASE tools can be rejuvenated when required by altering the systems specifications, already incorporated in the previous use of CASE, to take account of new needs. These can then be fed into back-end CASE tools to produce new code. The process is considerably quicker than attempting alterations via manually held systems specifications. Second, by the use of reverse engineering tools existing applications can be recast in a way that makes them suitable for amendment.

2. *Complete, accurate and consistent design specifications:* Most errors in the development of a new system occur in the systems specification stage. Most of these are not picked up until acceptance testing by users has taken place. CASE, by the use of computer specification tools and a central information repository, forces certain aspects of consistency and completeness on the design process.

3. *Reducing human effort:* CASE tools reduce human effort in analysis and design by offloading development work onto computers.

    Diagramming and charting tools cut the development time, especially when it is considered that each diagram may have to undergo several versions before completion.

    By keeping track of the development process CASE tools can relieve the human of considerable project management burdens. These tools keep note of authors, versions of models and a calendar. CASE tools also provide for consistency and completeness checking across various stages of the development process. They do this by tracking entities, process, data definitions, diagrams and other things that would otherwise take up much human effort.

    Back-end CASE tools significantly reduce the human programming requirement by generating source and object code, and database definitions.

4. *Integration of development:* CASE tools, particularly I-CASE tools, encourage integration of the development process from the early stages of corporate analysis through information systems specification to code implementation. This is a trend that has been emerging in paper-based methodologies and it is now mirrored in CASE tools and methodologies. It is based on a recognition that the design of an information system is not merely a technical issue deriving from a clear understanding of where we are now and where we want to go, but rather an activity that has to link and translate corporate information systems requirements into deliverable systems.

5. *Speed:* CASE tools speed the development process for a project. The tools also allow considerable interactive input from systems developers during the process. This is compatible with a prototyping approach to systems development.

6. *Links to object-oriented analysis:* Object-oriented analysis and design is fast becoming the dominant design approach of the 1990s. Object-oriented methods are particularly suited to incorporation in CASE tools. Central to any CASE approach is a central information repository, which can be viewed as information on objects.

OBJECT-ORIENTED ANALYSIS AND DESIGN

Data analysis and structured process analysis have been highly influential in generating commercial methodologies centered around one or both approaches. However, some of the difficulties experienced in analysis and design using these perspectives are rooted in the assumption that an organization can be understood separately in terms of processes that manipulate data and in terms of objects on which data is stored. This separation between processes and objects, as practitioners of object-oriented analysis and design would claim, is ill founded.

It is suggested that our understanding of the world is based on objects that have properties, stand in relationships to other objects, and take part in operations. For example, we understand a motor car in terms not just of its static properties and relationships (is a Ford, has four wheels) but also in terms of the operations in which it takes part (acceleration, transportation of people). In a business context an order is not merely an object that has an order number, date, an item number ordered, a quantity ordered, but also takes part in operations—is placed by a customer, generates a stock requisition. This notion of object is central to an understanding of object-oriented analysis.

From the perspective of programming rather than analysis, objects and object types were a natural development out of the need to define types of things together with permissible operations performable on them. In early programming languages there were only a few types of object (data types). Examples were integer and string. Certain operations were permissible on integers, such as addition, but not on strings. Other operations were defined for strings, such as concatenation, but not for integers. Later the ability to define many different data types, together with their permissible operations, became incorporated in languages. This meshed with an object-oriented approach to analysis.

■ ESSENTIAL CONCEPTS

OBJECT   An **object** is any thing on which we store data together with the operations that are used to alter that data. An **object type** is a category of object. For instance an object might be Bill Smith, which is an instance of the object type *employee*. Another object is order#1234, which is an instance of the object type *order*.

A **method** specifies the way in which an object's data is manipulated—the way in which an **operation** is carried out. For instance, the sales tax on an invoice might be produced from a method that takes the sales amount and calculates a percentage of this. The important point is that the object consists not only of the data but also of the operations that manipulate it. This is one of the important features that distinguishes an object from an entity as used in entity-relationship modelling.

ENCAPSULATION   The data held on an object is hidden from the user of that object. How then is data retrieved? All data operations and access to data are handled through methods. Packaging the data so that the data is hidden from users is known as **encapsulation.** The data is also hidden from other objects.

In implementation the methods are not considered to be packaged with the object but rather with the object type, as the same method for handling a certain type of data applies to all objects of that type. This corresponds well with object-oriented languages, which store the method as program code with the object type (or object class **as it is known).**

MESSAGE   In order to retrieve data, or indeed carry out any operation on an object, it is necessary to send a **message** to the object. This message will contain the name of the object, the object type, the name of the operation and any other relevant parameters—for example, data values in the case of an update operation. When an operation is invoked by a message, that message may have come from another object or, more strictly, from an operation associated with that object. The first object is said to send a **request** to the second object. The request may result in a response if that is part of the purpose of the operation. This is the way objects communicate.

INHERITANCE   Object types may have subtypes and supertypes. The types *dog* has as subtypes *Alsatian* and *poodle*, and as supertypes *carnivore* and *mammal*. All Alsatians are dogs. All dogs are mammals and all dogs are carnivores. Rover has certain properties by virtue of being an Alsatian; Fifi by virtue of being a poodle. Alsatians are said to **inherit** their dog characteristics by virtue of those properties being had by their supertype—*dog*. Similarly for poodles. Dogs inherit their mammalian characteristics, such as suckling young, from their supertype—*mammal*.

An object type will have some operations and methods specific to it. Others it will inherit from its supertype. This makes sense from the point of view of implementation, as it is only necessary to store a method at its highest level of applicability avoiding unnecessary duplication. *Manager* and *clerical staff* are two object types with the common supertype *employee*. An operation for employee might be *hire*. This would be inherited by both *manager* and *clerical staff.*

IMPLEMENTATION

After maintenance, the implementation phase of the systems development life cycle is the most expensive and time-consuming phase of the entire life cycle. Implementation is expensive because so many people are involved in the process; it is time-consuming because of all the work that has to be completed during implementation. Physical design specifications must be turned into working computer code, the code must be tested until most of the errors have been detected and corrected, the system must be installed, user sites must be prepared for the new system, and users must come to rely on the new system rather than the existing one to get their work done.

Implementing a new information system into an organizational context is not a mechanical process. The organizational context has been shaped and reshaped by the people who work in the organization. The work habits, beliefs, inter-relationships, and personal goals of an organization's members all affect the implementation process. Although factors important to successful implementation have been identified, there are no sure recipes you can follow. During implementation, you must be attuned to key aspects of the organizational context, such as history, politics, and environmental demands—aspects that can contribute to implementation failures if ignored.

Here you will learn about the many activities that the implementation phase comprises. In this section, we discuss coding, testing, installation, documentation, user training, support for a system after it is installed, and implementation success. Our intent is not to teach you how to program and test systems—most of you have already learned about writing and testing programs in the courses you took before this one. Rather, this section shows you where coding and testing fit in the overall scheme of implementation and stresses the view of implementation as an organizational change process that is not always successful.

In addition, you will learn about providing documentation about the new system for the information systems personnel who will maintain the system and for the system's users. These same users must be trained to use what you have developed and installed in their workplace. Once training has ended and the system has become institutionalized, users will have questions about the system's implementation and how to use it effectively. You must provide a means for users to get answers to these questions and to identify needs for further training.

As a member of the system development team that developed and implemented the new system, your job is winding down now that installation and conversion are complete. The end of implementation also marks the time for you to begin the process of project close-down.

After a brief overview of the coding, testing, and installation processes and the deliverables and outcomes from these processes, we will talk about software application testing. The following section presents the four types of installation: direct, parallel, single location, and phased. Afterwards, you will read about the process of documenting systems and training and supporting users as well as the deliverables from these processes. In the next section you will learn about the various types of documentation and numerous methods available for delivering training and support services. Finally, you will read about implementation as an organizational change process, with many organizational and people issues involved in the implementation effort.

SYSTEM IMPLEMENTATION

System implementation is made up of many activities. The six major activities we are concerned with in this section are coding, testing, installation, documentation, training, and support (see Figure 4-1). The purpose of these steps is to convert the final physical system specifications into working and reliable software and hardware, document the work that has been done, and provide help for current and future users and caretakers of the system. These steps are often done by other project team members besides analysts, although analysts may do some programming. In any case, analysts are responsible for ensuring that all of these various activities are properly planned and executed. Next, we will briefly discuss these activities in two groups, first coding, testing, and installation, and then, documenting the system and training and supporting users.

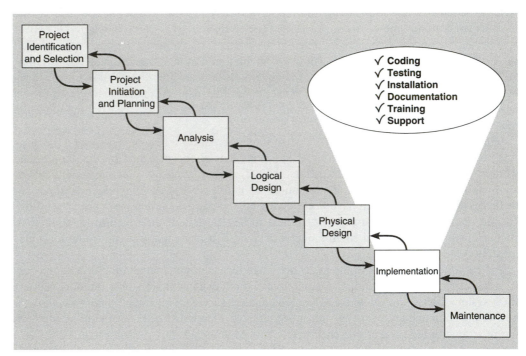

FIGURE 4-1   *The Systems Development Life Cycle with the Implementation Phase Highlighted*

■ THE PROCESS OF CODING, TESTING, AND INSTALLATION

Coding, as we mentioned before, is the process whereby the physical design specifications created by the analysis team are turned into working computer code by the programming team. Depending on the size and complexity of the system, coding can be an involved, intensive activity. Once coding has begun, the testing process can begin and proceed in parallel. As each program module is produced, it can be tested individually, then as part of a larger program, and then as part of a larger system. You will learn about the different strategies for testing later in the section. We should emphasize that although testing is done during implementation, you must begin planning for testing earlier in the project. Planning involves determining what needs to be tested and collecting test data. This is often done during the analysis phase because testing requirements are related to system requirements.

Installation is the process during which the current system is replaced by the new system. This includes conversion of existing data, software, documentation, and work procedures to those consistent with the new system. Users must give up the old ways of doing their jobs, whether manual or automated, and adjust to accomplishing the same tasks with the new system. Users will sometimes resist these changes and you must help them adjust. However, you cannot control all the dynamics of user-system interaction involved in the installation process.

■ DELIVERABLES AND OUTCOMES FROM CODING, TESTING, AND INSTALLATION

Table 4.1 shows the deliverables from the coding, testing, and installation processes. The most obvious outcome is the code itself, but just as important as the code is documentation of the code. Some programming languages, such as COBOL, are said to be largely self-documenting because the language itself spells out much about the program's logic, the labels used for data and variables, and the locations where data are accessed and output. But even COBOL code can be mysterious to maintenance programmers who must maintain the system for years after the original system was written and the original programmers have moved on to other jobs. Therefore, clear, complete documentation for all individual modules and programs is crucial to the system's continued smooth operation. Increasingly, CASE tools are used to maintain the documentation needed by systems professionals. The results of program and system testing are important deliverables from the testing process, as they document the tests as well as the test results. For example, what type of test was conducted? What test data were used? How did the system handle the test? The answers to these questions can provide important information for system maintenance as changes will require retesting and similar testing procedures will be used during the maintenance process.

TABLE 4.1
Deliverables for Coding, Testing, and Installation

1. Coding
   a. Code
   b. Program documentation
2. Testing
   a. Test scenarios (test plan) and test data
   b. Results of program and system testing
3. Installation
   a. User guides
   b. User training plan
   c. Installation and conversion plan
      i. Software and hardware installation-schedule
      ii. Data conversion plan
      iii. Site and facility remodeling plan

The next two deliverables, user guides and the user training plan, result from the installation process. User guides provide information on how to use the new system, and the training plan is a strategy for training users so they can quickly learn the new system. The development of the training plan probably began earlier in the project and some training, on the concepts behind the new system, may have already taken place. During the early stages of implementation, the training plans are finalized and training on the use of the system begins. Similarly, the installation plan lays out a strategy for moving from the old system to the new, from the beginning to the end of the process. Installation includes installing the system (hardware and software) at central and user sites. The installation plan answers such questions as when the new system will be installed, which installation strategies will be used, who will be involved, what resources are required, which data will be converted and cleansed, and how long the installation process will take. It is not enough that the system is installed; users must actually use it.

As an analyst, your job is to ensure that all of these deliverables are produced and are done well. You may produce some of the deliverables, such as test data, user guides, and an installation plan; for other deliverables, such as code, you may only supervise or simply monitor their production or accomplishment. The extent of your implementation responsibilities will vary according to the size and standards of the organization you work for, but your ultimate role includes ensuring that all the implementation work leads to a system that meets the specifications developed in earlier project phases. See the box, The Future Programmer, for some insights into the changing role of programming and the nature of systems implementation in systems development.

■ THE PROCESSES OF DOCUMENTING THE SYSTEM, TRAINING USERS, AND SUPPORTING USERS

Although the process of documentation proceeds throughout the life cycle, it receives formal attention during the implementation phase because the end of implementation largely marks the end of the analysis team's involvement in system development. As the team is getting ready to move on to new projects, you and the other analysts need to prepare documents that reveal all of the important information you have learned about this system during its development and implementation. There are two audiences for this final documentation: (1) the information systems personnel who will maintain the system throughout its productive life and (2) the people who will use the system as part of their daily lives. The analysis team in a large organization can get help in preparing documentation from specialized staff in the information systems department.

Larger organizations also tend to provide training and support to computer users throughout the organization. Some of the training and support is very specific to particular application systems while the rest is general to particular operating systems or off-the-shelf software packages. For example, it is common to find courses on Microsoft Windows™ and WordPerfect™ in organization-wide training facilities. Analysts are mostly uninvolved with general training and support, but they do work with corporate trainers to provide training and support tailored to particular computer applications they have helped develop. Centralized information system training facilities tend to have specialized staff who can help with training and support issues. In smaller organizations that cannot afford to have well-staffed centralized training and support facilities, fellow users are the best source of training and support users have, whether the software is customized or off-the-shelf (Nelson and Cheney, 1987).

The Future Programmer

Where and by whom programming is done continues to change as the nature of programming languages evolves, resulting in improved programming productivity and the opening of programming to less highly skilled personnel. One prediction suggests that future programmers can be grouped into four categories:

- IS department programmers: the number of people who work in IS are in a clear decline, from nearly 2 million in 1994 to several hundred thousand by 2010; some people believe that these jobs are really being distributed out of the central IS function into business units, possibly under different titles.

- Software company programmers: These programmers work for consulting and packaged software companies, and the number will likely rise from roughly 600,000 in 1994 to several million by 2010.

- Embedded software programmers: These programmers produce code that is embedded in other products, like cars, office equipment, and consumer electronics; this group will likely dramatically increase from several million in 1994 to over 10 million by 2010.

- Occasional programmers: Professionals and technicians (accountants, engineers, managers, and so forth) who program as part of their main duties; this group should rise from roughly 20 million in 1994 to over 100 million by 2010.

One theory is that standard business system components (objects in some terminologies) can be assembled into new systems by less skilled programmers. Thus, the job of what we call today a programmer will be to build these components and to ensure the quality of assembled systems. Although the number of occasional programmers (likely not trained in information systems) is exploding, the need for highly skilled programmers and programming work is far from diminishing.

Adapted from Bloor, 1994

■ DELIVERABLES AND OUTCOMES FROM DOCUMENTING THE SYSTEM, TRAINING USERS, AND SUPPORTING USERS

Table 4.2 shows the deliverables from documenting the system, training users, and supporting users. At the very least, the development team must prepare user documentation. The documentation can be paper-based, but it should also include computer-

TABLE 4.2
Deliverables for Documenting the System, Training, and Supporting Users

1. Documentation
   a. System documentation
   b. User documentation
2. User training plan
   a. Classes
   b. Tutorials
3. User training modules
   a. Training materials
   b. Computer-based training aids
4. User support plan
   a. Help desk
   b. On-line help
   c. Bulletin boards and other support mechanisms

based modules. For modern information systems, this documentation includes any on-line help designed as part of the system interface. The development team should think through the user training process: Who should be trained? How much training is adequate for each training audience? What do different types of users need to learn during training? The training plan should be supplemented by actual training modules, or at least outlines of such modules, that at a minimum address the three questions stated previously. Finally, the development team should also deliver a user support plan that addresses such issues as how users will be able to find help once the information system has become integrated into the organization. The development team should consider a multitude of support mechanisms and modes of delivery.

SOFTWARE APPLICATION TESTING

As we mentioned previously, analysts prepare system specifications that are passed on to programmers for coding. Although coding takes considerable effort and skill, the practices and processes of writing code do not belong in this text. However, as software application testing is an activity analysts plan (beginning in the analysis phase) and sometimes supervise, depending on organizational standards, you need to understand the essentials of the testing process.

Testing software begins earlier in the systems development life cycle, even though many of the actual testing activities are carried out during implementation. During analysis, you develop a master test plan. During design, you develop a unit test plan, an integration test plan, and a system test plan. During implementation, these various plans are put into effect and the actual testing is performed.

The purpose of these written test plans is to improve communication among all the people involved in testing the application software. The plan specifies what each person's role will be during testing. The test plans also serve as checklists you can use to determine whether all of the master test plan has been completed. The master test plan is not just a single document but a collection of documents. Each of the component documents represents a complete test plan for one part of the system or for a particular type of test. Presenting a complete master test plan is far beyond the scope of this book. Refer to Mosley's *Handbook of MIS Software Application Testing* for a complete test plan, which comprises a 101-page appendix. To give you an idea of what a master test plan involves, we present an abbreviated table of contents in Table 4.3 on the next page.

A master test plan is a project within the overall system development project. Since at least some of the system testing will be done by people who have not been involved in the system development so far, the Introduction provides general information about the system and the needs for testing. The Overall Plan and Testing Requirements sections are like a baseline project plan for testing, with a schedule of events, resource requirements, and standards of practice outlined. Procedure Control explains how the testing is to be conducted, including

TABLE 4.3
Table of Contents of a Master Test Plan

1. Introduction
    a. Description of system to be tested
    b. Objectives of the test plan
    c. Method of testing
    d. Supporting documents

2. Overall Plan
    a. Milestones, schedule, and locations
    b. Test materials
        1. Test plans
        2. Test cases
        3. Test scenarios
        4. Test log
    c. Criteria for passing tests

3. Testing Requirements
    a. Hardware
    b. Software
    c. Personnel

4. Procedure Control
    a. Test initiation
    b. Test execution
    c. Test failure
    d. Access/change control
    e. Document control

5. Test Specific or Component Specific Test Plans
    a. Objectives
    b. Software description
    c. Method
    d. Milestones, schedule, progression, and locations
    e. Requirements
    f. Criteria for passing tests
    g. Resulting test materials
    h. Execution control
    i. Attachments

Adapted from Mosley, 1993

how changes to fix errors will be documented. The fifth and final section explains each specific test necessary to validate that the system performs as expected.

Some organizations have specially trained personnel who supervise and support testing. Testing managers are responsible for developing test plans, establishing testing standards, integrating testing and development activities in the life cycle, and ensuring that test plans are completed. Testing specialists help develop test plans, create test cases and scenarios, execute the actual tests, and analyze and report test results.

■ SEVEN DIFFERENT TYPES OF TESTS

Software application testing is an umbrella term that covers several types of tests. Mosley (1993) organizes the types of tests according to whether they employ static or dynamic techniques and whether the test is automated or manual. Static testing means that the code being tested is not executed. The results of running the code are not an issue for that particular test. Dynamic testing, on the other hand, involves execution of the code. Automated testing means the computer conducts the test while manual means that people do. Using this framework, we can categorize types of tests as shown in Table 4.4.

Let's examine each type of test in turn. Inspections are formal group activities where participants manually examine code for occurrences of well-known errors. Syntax, grammar, and some other routine errors can be checked by automated inspection software, so manual inspection checks are used for more subtle errors. Each programming language lends itself

TABLE 4.4
A Categorization of Test Types

	Manual	Automated
Static	Inspections	Syntax checking
Dynamic	Walkthroughs	Unit test
	Desk checking	Integration test
		System test

Adapted from Mosley, 1993

to certain types of errors that programmers make when coding, and these common errors are well-known and documented (for common coding errors in COBOL, see Litecky and Davis, 1976). Code inspection participants compare the code they are examining to a checklist of well-known errors for that particular language. Exactly what the code does is not investigated in an inspection. It has been estimated that code inspections have been used by organizations to detect from 60 to 90 percent of all software defects as well as to provide programmers with feedback that enables them to avoid making the same types of errors in future work (Fagan, 1986). The inspection process can also be used for such things as design specifications.

Unlike inspections, what the code does is an important question in a *walkthrough*. Using structured walkthroughs is a very effective method of detecting errors in code. Structured walkthroughs can be used to review many systems development deliverables, including logical and physical design specifications as well as code. Whereas specification walkthroughs tend to be formal reviews, code walkthroughs tend to be informal. Informality tends to make programmers less apprehensive about walkthroughs and helps increase their frequency. According to Yourdon (1989), code walkthroughs should be done frequently when the pieces of work reviewed are relatively small and before the work is formally tested. If walkthroughs are not held until the entire program is tested, the programmer will have already spent too much time looking for errors that the programming team could have found much more quickly. The programmer's time will have been wasted, and the other members of the team may become frustrated because they will not find as many errors as they would have if the walkthrough had been conducted earlier. Further, the longer a program goes without being subjected to a walkthrough, the more defensive the programmer becomes when the code is reviewed. Although each organization that uses walkthroughs conducts them differently, there is a basic structure that you can follow that works well (see Figure 4-2).

It should be stressed that the purpose of a walkthrough is to detect errors, not to correct them. It is the programmer's job to correct the errors uncovered in a walkthrough. Sometimes it can be difficult for the reviewers to refrain from suggesting ways to fix the problems they find in the code, but increased experience with the process can help change a reviewer's behavior.

What the code does is also important in desk checking, an informal process where the programmer or someone else who understands the logic of the program works through the code with a paper and pencil. The programmer executes each instruction, using test cases that may or may not be written down. In one sense, the reviewer acts as the computer, mentally checking each step and its results for the entire set of computer instructions.

Among the list of automated testing techniques in Table 4.4, there is only one technique that is also static, syntax checking. Syntax checking is typically done by a compiler. Errors in syntax are uncovered but the code is not executed. For the other three automated techniques, the code is executed.

GUIDELINES FOR CONDUCTING A CODE WALKTHROUGH

1. Have the review meeting chaired by the project manager or chief programmer, who is also responsible for scheduling the meeting, reserving a room, setting the agenda, inviting participants, and so on.
2. The programmer presents his or her work to the reviewers. Discussion should be general during the presentation.
3. Following the general discussion, the programmer walks through the code in detail, focusing on the logic of the code rather than on specific test cases.
4. Reviewers ask to walk through specific test cases.
5. The chair resolves disagreements if the review team cannot reach agreement among themselves and assigns duties, usually to the programmer, for making specific changes.
6. A second walkthrough is then scheduled if needed.

FIGURE 4-2   *Steps in a Typical Walkthrough*

The first such technique is unit testing, sometimes called module testing. In unit testing, each module is tested alone in an attempt to discover any errors that may exist in the module's code. But since modules coexist and work with other modules in programs and systems, they must be tested together in larger groups. Combining modules and testing them is called integration testing. Integration testing is gradual. First you test the coordinating module (the root module in a structure chart tree) and only one of its subordinate modules. After the first test, you add one or two other subordinate modules from the same level. Once the program has been tested with the coordinating module and all of its immediately subordinate modules, you add modules from the next level and then test the program. You continue this procedure until the entire program has been tested as a unit. System testing is a similar process, but instead of integrating modules into programs for testing, you integrate programs into systems. System testing follows the same incremental logic that integration testing does. Under both integration and system testing, not only do individual modules and programs get tested many times, so do the interfaces between modules and programs.

Current practice calls for a top-down approach to writing and testing modules. Under a top-down approach, the coordinating module is written first. Then the modules at the next level in the structure chart are written, followed by the modules at the next level, and so on, until all of the modules in the system are done. Each module is tested as it is written. Since top-level modules contain many calls to subordinate modules, you may wonder how they can be tested if the lower level modules haven't been written yet. The answer is stub testing. Stubs are two or three lines of code written by a programmer to stand in for the missing modules. During testing, the coordinating module calls the stub instead of the subordinate module. The stub accepts control and then returns it to the coordinating module.

Figure 4-3 illustrates stub and integration system testing. Stub testing is depicted as the innermost oval. Here the Get module (where data are input and read) is being written and tested, but as none of its subordinate modules have been written yet, each one is represented by a stub. In the stub testing illustrated by Figure 4-3, Get is tested with only one stub in place, for its left-most subordinate module. You would of course write stubs for all of the Get module's subordinate modules, just as you would for the Make (where new information is calculated) and Put (where information is output) modules. Once all of the subordinate modules are written and tested, you would conduct an integration test of Get and its subordinate modules, as represented by the second oval. As stated previously, the focus of an integration test is on the inter-relationships among modules. You would also conduct integration tests of

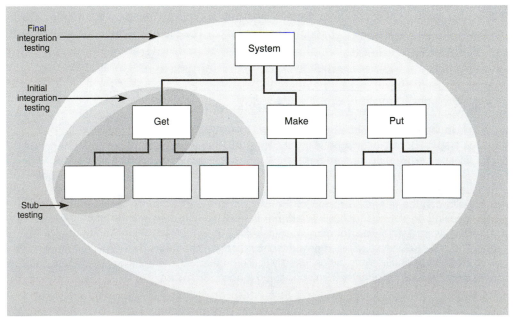

FIGURE 4-3   *Comparing Stub and Integration Testing*

Make and its subordinate modules, of Put and its subordinates, and of System and its subordinates, Get, Make, and Put. Eventually your integration testing would include all of the modules in the large oval that encompasses the entire program.

System testing is more than simply expanded integration testing where you are testing the interfaces between programs in a system rather than testing the interfaces between modules in a program. System testing is also intended to demonstrate whether a system meets its objectives. This is not the same as testing a system to determine whether it meets requirements—that is the focus of acceptance testing, which will be discussed later. To verify that a system meets its objectives, system testing involves using nonlive test data in a nonlive testing environment. *Nonlive* means the data and situation are artificial, developed specifically for testing purposes, although both data and environment are similar to what users would encounter in everyday system use. The system test is typically conducted by information systems personnel led by the project team leader, although it can also be conducted by users under MIS guidance. The scenarios that form the basis for system tests are prepared as part of the master test plan.

■ THE TESTING PROCESS

Up to this point, we have talked about the master test plan and seven different types of tests for software applications. We haven't said very much about the process of testing itself. There are two important things to remember about testing information systems:

1. The purpose of testing is confirming that the system satisfies requirements.
2. Testing must be planned.

These two points have several implications for the testing process, regardless of the type of test being conducted. First, testing is not haphazard. You must pay attention to many different aspects of a system, such as response time, response to boundary data, response to no input, response to heavy volumes of input, and so on. You must test anything (within resource constraints) that could go wrong or be wrong about a system. At a minimum, you should test the most frequently used parts of the system and as many other paths through the system as time permits. Planning gives analysts and programmers an opportunity to think through all the potential problem areas, list these areas, and develop ways to test for problems. As indicated previously, one part of the master test plan is creating a set of test cases, each of which must be carefully documented (see Figure 4-4 for an outline of a test case description).

A test case is a specific scenario of transactions, queries, or navigation paths that represent a typical, critical, or abnormal use of the system. A test case should be repeatable, so that it can be rerun as new versions of the software are tested. This is important for all code, whether written in-house, developed by a contractor, or purchased. Test cases need to determine that new software works with other existing software with which it must share data. Even though analysts often do not do the testing, systems analysts, because of their intimate knowledge of applications, often make up or find test data. The people who create the test cases should not be the same people as those who coded and tested the system. In addition to a description of each test case, there must also be a description of the test results, with an emphasis on how the actual results differed from the expected results (see Figure 4-5). This description will indicate why the results were different and what, if anything, should be done to change the software. This description will then suggest the need for retesting, possibly introducing new tests necessary to discover the source of the differences.

One important reason to keep such a thorough description of test cases and results is so that testing can be repeated for each revision of an application. Although new versions of a system may necessitate new test data to validate new features of the application, previous test data usually can and should be reused. Results from use of the test data with prior versions are compared to new versions to show that changes have not introduced new errors and that the behavior of the system, including response time, is no worse. A second implication for the testing process is that test cases must include illegal and out-of-range data. The system should be able to handle any possibility, no matter how unlikely; the only way to find out is to test.

```
Pine Valley Furniture Company
Test Case Description

Test Case Number:
Date:
Test Case Description:



Program Name:
Testing State:
Test Case Prepared By:

Test Administrator:

Description of Test Data:



Expected Results:



Actual Results:
```

FIGURE 4-4   *Test Case Description Form
(Adapted from Mosley, 1993)*

```
Pine Valley Furniture Company
Test Case Results

Test Case Number:
Date:

Program Name:
Module Under Test:

Explanation of difference between actual and expected output:






Suggestions for next steps:
```

FIGURE 4-5   *Test Results Form (Adapted from
Mosley, 1993)*

If the results of a test case do not compare favorably to what was expected, the error causing the problem must be found and fixed. Programmers use a variety of debugging tools to help locate and fix errors. A sophisticated debugging tool called a *symbolic debugger* allows the program to be run on-line, even one instruction at a time if the programmer desires, and allows the programmer to observe how different areas of data are affected as the instructions are executed. This cycle of finding problems, fixing errors, and rerunning test cases continues until no additional problems are found. There are specific testing methods that have been developed for generating test cases and guiding the test process (Mosley, 1993). See the box, Automating Testing, for an overview of tools to assist you in testing software.

Automating Testing

Software testing tools provide the following functions, which improve the quality of testing:

- Record or build scripts of data entry, menu selections and mouse clicks, and input data, which can be replayed in exact sequence for each test run

- Compare the results of one test run with those from prior test cases to identify errors or to highlight the results of new features

- Supported unattended script playing to simulate high-volume or stress situations. Such tools can reduce the time for software testing by almost 80 percent.

■ ACCEPTANCE TESTING BY USERS

Once the system tests have been satisfactorily completed, the system is ready for acceptance testing, which is testing the system in the environment where it will eventually be used. Acceptance refers to the fact that users typically sign off on the system and "accept" it once they are satisfied with it. As we said previously, the purpose of acceptance testing is for users to determine whether the system meets their requirements. The extent of acceptance testing will vary with the organization and with the system in question. The most complete acceptance testing will include alpha testing, where simulated but typical data are used for system testing; beta_testing, in which live data are used in the users' real working environment; and a system audit conducted by the organization's internal auditors or by members of the quality assurance group.

During alpha testing, the entire system is implemented in a test environment to discover whether or not the system is overtly destructive to itself or to the rest of the environment. The types of tests performed during alpha testing include the following:

- Recovery testing—forces the software (or environment) to fail in order to verify that recovery is properly performed.

- Security testing—verifies that protection mechanisms built into the system will protect it from improper penetration.

- Stress testing—tries to break the system (for example, what happens when a record is written to the database with incomplete information or what happens under extreme on-line transaction loads or with a large number of concurrent users).

- Performance testing—determines how the system performs on the range of possible environments in which it may be used (for example, different hardware configurations, networks, operating systems, and so on); often the goal is to have the system perform with similar response time and other performance measures in each environment.

In beta_testing, a subset of the intended users run the system in their own environments using their own data. The intent of the beta test is to determine whether the software, documentation, technical support, and training activities work as intended. In essence, beta testing can be viewed as a rehearsal of the installation phase. Problems uncovered in alpha and beta testing in any of these areas must be corrected before users can accept the system. There are many stories systems analysts can tell about long delays in final user acceptance due to system bugs (see the box, Bugs in the Baggage, for one famous incident).

INSTALLATION

The process of moving from the current information system to the new one is called installation. All employees who use a system, whether they were consulted during the development process or not, must give up their reliance on the current system and begin to rely on the new system. Four different approaches to installation have emerged over the years: direct, parallel, single location, and phased (Figure 4-6). The approach an organization decides to use will depend on the scope and complexity of the change associated with the new system and the organization's risk aversion.

■ DIRECT INSTALLATION

The direct or abrupt approach to installation (also called "cold-turkey") is as sudden as the name indicates: the old system is turned off and the new system is turned on (Figure 4-6a). Under direct installation, users are at the mercy of the new system. Any errors resulting from the new system will have a direct impact on the users and how they do their jobs and, in some cases—depending on the centrality of the system to the organization—on how the organization performs its business. If the new system fails,

Bugs in the Baggage

Testing a complex software system can be long and frustrating. A case in point was the software to control 4,000 baggage cars at the Denver International Airport. Errors in the software put the airport's opening on hold for months, costing taxpayers $500,000 a day and turning airport bonds into junk status. The airport was supposed to have opened in March 1994, but it did not open until February 1995 because of problems with the baggage-handling system. The system routinely damaged luggage and routed bags to the wrong flights. Various causes of the delay were identified, including last minute design change requests from airport officials and mechanical problems. The bottom-line lesson is that system designers must build in plenty of test and debugging time when scaling up proven technology into a much more complicated environment.

United Airlines, the major carrier at Denver International Airport, took over as systems integrator in October 1994. At the same time, the City of Denver commissioned a traditional conveyor-belt baggage-handling system in the airport for an additional $50 million. When the airport opened in 1995 (and as of this writing), only United used the automated baggage system, and then only to ferry bags to its flights. The system was not able to ferry bags from the planes back to the airport. All the other carriers used the traditional conveyor system. The City of Denver tried to recover $80 million of the system's $193 million cost from BAE Automated Systems Inc., the system's vendor. In 1996, BAE sued United for $17.5 million and the city for $4.1 million in withheld fees. United countersued. In September 1997, all sides settled, and no details were released.

Sources include Bozman, 1994; Lieb, 1997; and Scheier, 1994.

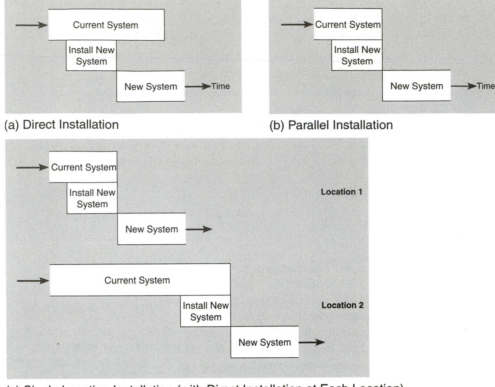

(a) Direct Installation   (b) Parallel Installation

(c) Single Location Installation (with Direct Installation at Each Location)

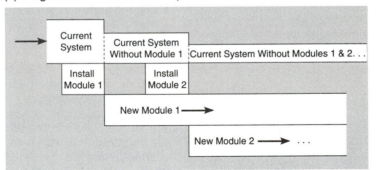

(d) Phased Installation

FIGURE 4-6   *Comparison of Installation Strategies*

considerable delay may occur until the old system can again be made operational and business transactions reentered to make the database up-to-date. For these reasons, direct installation can be very risky. Further, direct installation requires a complete installation of the whole system. For a large system, this may mean a long time until the new system can be installed, thus delaying system benefits or even missing the opportunities that motivated the system request. On the other hand, it is the least expensive installation method and it creates considerable interest in making the installation a success. Sometimes, a direct installation is the only possible strategy if there is no way for the current and new systems to coexist, which they must do in some way in each of the other installation approaches.

■ PARALLEL INSTALLATION

Parallel installation is as riskless as direct installation is risky. Under parallel installation, the old system continues to run alongside the new system until users and management are satisfied that the new system is effectively performing its duties and the old system can be turned off (Figure 4-6b). All of the work done by the old system is concurrently performed by the new system. Outputs are compared (to the greatest extent possible) to help determine whether the new system is performing as well as the old. Errors discovered in the new system do not cost the organization much, if anything, as errors can be isolated and the business can be supported with the old system. Since all work is essentially done twice, a parallel installation can be very expensive; running two systems implies employing (and paying) two staffs to not only operate both systems but also to maintain them. A parallel approach can also be confusing to users since they must deal with both systems. As with direct installation, there can be a considerable delay until the new system is completely ready for installation. A parallel approach may not be feasible, especially if the users of the system (such as customers) cannot tolerate redundant effort or the size of the system (number of users or extent of features) is large.

■ SINGLE LOCATION INSTALLATION

Single location installation, also known as location and pilot installation, is a middle-of-the-road approach compared to direct and parallel installation. Rather than convert all of the organization at once, single location installation involves changing from the current to the new system in only one place or in a series of separate sites over time (Figure 4-6c depicts this approach for a simple situation of two locations). The single location may be a branch office, a single factory, or one department, and the actual approach used for installation in that location may be any of the other approaches. The key advantage to single location installation is that it limits potential damage and potential cost by limiting the effects to a single site. Once management has determined that installation has been successful in one location, the new system may be deployed in the rest of the organization, possibly continuing with installation in one location at a time. Success at the pilot site can be used to convince reluctant personnel in other sites that the system can be worthwhile for them as well. Problems with the system (the actual software as well as documentation, training, and support) can be resolved before deployment to other sites. Even though the single location approach may be simpler for users, it still places a large burden on IS staff to support two versions of the system. On the other hand, because problems are isolated at one site at a time, IS staff can devote all its efforts to the success at the pilot site. Also, if different locations require sharing of data, extra programs will need to be written to synchronize the current and new systems; although this will happen transparently to users, it is extra work for IS staff. As with each of the other approaches (except phased installation), the whole system is installed; however, some parts of the organization will not get the benefits of the new system until the pilot installation has been completely tested.

■ PHASED INSTALLATION

Phased installation, also called staged installation, is an incremental approach. Under phased installation, the new system is brought on-line in functional components; different parts of the old and new systems are used in cooperation until the whole new system is installed (Figure 4-6d shows the phase-in of the first two modules of a new system). Phased installation, like single location installation, is an attempt to limit the organization's exposure to risk, whether in terms of cost or disruption of the business. By converting gradually, the organization's risk is spread out over time and place. Also, a phased installation allows for some benefits from the new system before the whole system is ready. For example, new data-capture methods can be used before all reporting modules are ready. For a phased installation, the new and replaced systems must be able to coexist and probably share data. Thus, bridge programs connecting old and new databases and programs often must be built. Sometimes, the new and old systems are so incompatible (built using totally different structures) that pieces of the old system cannot be incrementally replaced, so this strategy is not feasible. A phased installation is akin to bringing out a sequence of releases of the system. Thus, a phased approach requires careful version control, repeated conversions at each phase, and a long period of change, which may be frustrating and confusing to users. On the other hand, each phase of change is smaller and more manageable for all involved.

■ PLANNING INSTALLATION

Each installation strategy involves converting not only software but also data and (potentially) hardware, documentation, work methods, job descriptions, offices and other facilities, training materials, business forms, and other aspects of the system. For example, it is necessary to recall or replace all the current system documentation and business forms, which suggests that the IS department must keep track of who has these items so that they can be notified and receive replacement items. In practice you will rarely choose a single strategy to the exclusion of all others; most installations will rely on a combination of two or more approaches. For example, if you choose a single location strategy, you have to decide how installation will proceed there and at subsequent sites. Will it be direct, parallel, or phased?

Of special interest in the installation process is the conversion of data. Since existing systems usually contain data required by the new system, current data must be made error-free, unloaded from current files, combined with new data, and loaded into new files. Data may need to be reformatted to be consistent with more advanced data types supported by newer technology used to build the new system. New data fields may have to be entered in large quantities so that every record copied from the current system has all the new fields populated. Manual tasks, such as taking a physical inventory, may need to be done in order to validate data before they are transferred to the new files. The total data conversion process can be tedious. Furthermore, this process may require that current systems be shut off while the data are extracted so that updates to old data, which would contaminate the extract process, cannot occur.

Any decision that requires the current system to be shut down, in whole or in part, before the replacement system is in place must be done with care. Typically, off hours are used for installations that require a lapse in system support. Whether a lapse in service is required or not, the installation schedule should be announced to users well in advance to let them plan their work schedules around outages in service and periods when their system support might be erratic. Successful installation steps should also be announced, and special procedures put in place so that users can easily inform you of problems they encounter during installation periods. You should also plan for emergency staff to be available in case of system failure so that business operations can be recovered and made operational as quickly as possible. Another consideration is the business cycle of the organization. Most organizations face heavy workloads at particular times of year and relatively light loads at other times. A well-known example is the retail industry, where the busiest time of year is the fall, right before the year's major gift-giving holidays. You wouldn't want to schedule installation of a new point-of-sale system to begin December 1 for a department store. Make sure you understand the cyclical nature of the business you are working with before you schedule installation.

Planning for installation may begin as early as the analysis of the organization supported by the system. Some installation activities, such as buying new hardware, remodeling facilities, validating data to be transferred to the new system, and collecting new data to be loaded into the new system, must be done before the software installation can occur. Often the project team leader is responsible for anticipating all installation tasks and assigns responsibility for each to different analysts.

Each installation process involves getting workers to change the way they work. As such, installation should be looked at not as simply installing a new computer system, but as an organizational change process. More than just a computer system is involved—you are also changing how people do their jobs and how the organization operates.

DOCUMENTING THE SYSTEM

In one sense, every information systems development project is unique and will generate its own unique documentation. In another sense, though, system development projects are probably more alike than they are different. Each project shares a similar systems development life cycle, which dictates that certain activities be undertaken and each of those activities be documented. Bell and Evans (1989) illustrate how a generic systems development life cycle maps onto a generic list of when specific systems development documentation elements are finalized (Table 4.5). As you compare the generic life cycle in Table 4.5 to the life cycle presented in this book, you will see that there are differences, but the general structure of both life cycles is the same, as both include the basic phases of analysis, design, implementation, and project planning. Specific documentation will vary depending on the life cycle you are following, and the format and content of the documentation may be mandated by the organization you work for. However, a basic outline of documentation can be adapted for specific needs, as shown in Table 4.5. Note that this table indicates when documentation is typically finalized; you should start developing documentation elements early, as the information needed is captured.

TABLE 4.5
SDLC and Generic Documentation Corresponding to Each Phase

Generic life cycle phase	Generic document
Requirements specification	System requirements specification
	Resource requirements specification
Project control structuring	Management plan
	Engineering change proposal
System development	
Architectural design	Architecture design document
Prototype design	Prototype design document
Detailed design & implementation	Detailed design document
Test specification	Test specifications
Test implementation	Test reports
System delivery	User's guide
	Release description
	System administrator's guide
	Reference guide
	Acceptance sign-off

Adapted from Bell and Evans, 1989

We can simplify the situation even more by dividing documentation into two basic types, system documentation and user documentation. System documentation records detailed information about a system's design specifications, its internal workings, and its functionality. In Table 4.5, all of the documentation listed (except for System delivery) would qualify as system documentation. System documentation can be further divided into internal and external documentation (Martin and McClure, 1985). Internal documentation is part of the program source code or is generated at compile time. External documentation includes the outcome of all of the structured diagramming techniques you have studied in this book, such as data flow and entity-relationship diagrams. Although not part of the code itself, external documentation can provide useful information to the primary users of system documentation—maintenance programmers. For example, structure charts and Nassi-Shneiderman charts together provide a good overview of a system's larger structure and the details of its inner workings. Nassi-Shneiderman charts are themselves used to model specifications for writing code. In the past, external documentation was typically discarded after implementation, primarily because it was considered too costly to keep up-to-date, but today's CASE environment makes it possible to maintain and update external documentation as long as desired.

While system documentation is intended primarily for maintenance programmers, user documentation is intended primarily for users. An organization may have definitive standards on system documentation, often consistent with CASE tools and the system development process. These standards may include the outline for the project dictionary and specific pieces of documentation within it. Standards for user documentation are not as explicit.

■ USER DOCUMENTATION

User documentation consists of written or other visual information about an application system, how it works, and how to use it. An excerpt of on-line user documentation for Microsoft Access™ appears in Figure 4-7. Notice how the documentation has hot links to the meaning of important terms. The documentation lists the steps necessary to actually perform the task the user inquired about. The "notes" section that follows explains specific restrictions and constraints that will affect what the user is attempting. You should also notice how some words in the documentation are underlined with dotted lines and that icons are used to represent Access buttons. This notation signifies that these particular words and buttons are hypertext links to related material elsewhere in the documentation. Hypertext techniques, rare in on-line PC documentation five years ago, are now the rule rather than the exception.

Figure 4-7 represents the content of a reference guide, just one type of user documentation (there is also a quick reference guide). Other types of user documentation include a user's guide, release description, system administrator's guide, and acceptance sign-off (Table 4.5). The reference guide consists of an exhaustive list of the system's functions and commands, usually in alphabetical order. Most on-line reference guides allow you to search by topic area or by typing in the first few letters of your keyword. Reference guides are very good for very specific information (as in Figure 4-7) but not as good for the broader picture of how you perform all the steps required for a given task. The quick reference guide provides essential information about operating a system in a short, concise format. Where computer resources are shared and many users perform similar tasks on the same machines (as with airline reservation or mail order catalog clerks), quick reference guides are often print-

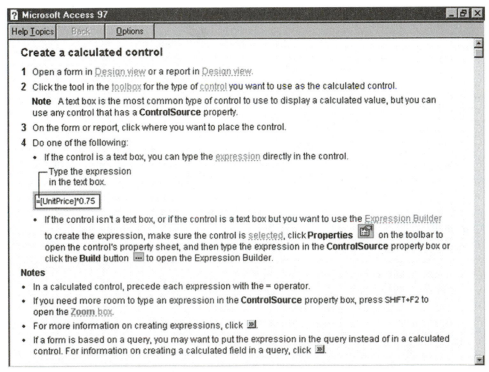

FIGURE 4-7   *Example of On-line User Documentation (From Microsoft Access™)*

ed on index cards or as small books and mounted on or near the computer terminal. An outline for a generic user's guide (from Bell and Evans, 1989) is shown in Table 4.6. The purpose of such a guide is to provide information on how users can use computer systems to perform specific tasks. The information in a user's guide is typically ordered by how often tasks are performed and how complex they are.

In Table 4.6, sections with an "n" and a title in square brackets mean that there would likely be many such sections, each for a different topic. For example, for an accounting application, sections 4 and beyond might address topics such as entering a transaction in the ledger, closing the month, and printing reports. The items in parentheses are optional, included as necessary. An index becomes more important the larger the user's guide. Although a generic user's guide outline is helpful in providing an overview for you of what a user's guide might contain, we have included outlines of user guides from several popular PC software packages in Figure 4-8. Notice how different they are.

TABLE 4.6
Outline of a Generic User's Guide

Preface

1. Introduction	3.3	Save
1.1 Configurations	3.4	Error recovery
1.2 Function flow	3.n	[Basic procedure name]
2. User interface	n.	[Task name]
2.1 Display screens	Appendix A—Error Messages	
2.2 Command types	([Appendix])	
3. Getting started	Glossary	
3.1 Login	Terms	
3.2 Logout	Acronyms	
	(Index)	

Adapted from Bell and Evans, 1989

A release description contains information about a new system release, including a list of complete documentation for the new release, features and enhancements, known problems and how they have been dealt with in the new release, and information about installation. A system administrator's guide is intended primarily for those who will install and administer a new system and contains information about the network on which the system will run, software interfaces for peripherals such as printers, troubleshooting, and setting up user accounts. Finally, the acceptance sign-off allows users to test for proper system installation and then signify their acceptance of the new system with their signatures.

ORGANIZATIONAL ISSUES IN SYSTEMS IMPLEMENTATION

Despite the best efforts of the systems development team to design and build a quality system and to manage the change process in the organization, the implementation effort sometimes fails. Sometimes employees will not use the new system that has been developed for them or, if they do use the system, their level of satisfaction with it is very low. Why do systems implementation efforts fail? This question has been the subject of information systems research for the past 25 years. In this section, we will try to provide some answers, first by looking at what conventional wisdom says are important factors related to implementation success, then by investigating factor-based models and political models of systems implementation.

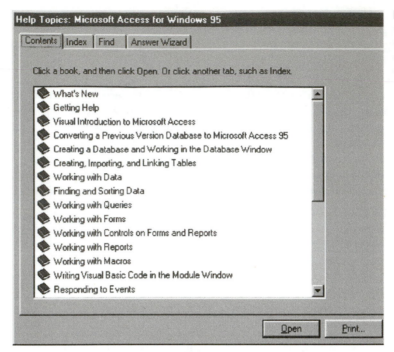

FIGURE 4-8
Outlines of User's Guides from Various Popular PC Software Packages

(a) Microsoft Access™

(b) WordPerfect™

■ WHY IMPLEMENTATION SOMETIMES FAILS

The conventional wisdom that has emerged over the years is that there are at least two conditions necessary for a successful implementation effort: management support of the system under development and the involvement of users in the development process (Ginzberg, 1981b). Conventional wisdom holds that if both of these conditions are met, you should have a successful implementation. Yet, despite the support and active participation of management and users, information systems implementation sometimes fails (see the box, System Implementation Failures, for examples).

System Implementation Failures

In 1985, the New Jersey Motor Vehicles Division implemented a new vehicle registration system. The new system had been developed by a major accounting firm in Applied Data Research's Ideal, a 4GL. Although appropriate for management information systems or decision support systems, 4GLs are not well-suited for high-volume transaction processing systems. As might have been predicted, the vehicle registration system was a disaster. Using the system entailed as much as a 5-minute response time, instead of the 2-second time the Division requested. System use resulted in a large number of incorrect vehicle registrations: over one million drivers in New Jersey had problems with their car registrations. The critical parts of the system had to be reprogrammed in COBOL at a cost of over two million dollars (which the accounting firm paid for) (Zwass, 1992).

Another state system has had an even more colorful history. In 1988, the Department of Health and Rehabilitative Services (HRS) began work on the development of the Florida On-Line Recipient Integrated Data Access system, or FLORIDA for short. The purpose of FLORIDA was to provide a single point of eligibility testing for the welfare services administered by the state, including food stamps, Aid to Families with Dependent Children (AFDC), and Medicaid. Access to FLORIDA would be possible from state welfare offices throughout the state, but FLORIDA and all of its files would reside in Tallahassee, the state capital. In 1989, EDS Federal Corporation won the bid for primary contractor for FLORIDA. One of the contract's stipulations was that the system be developed and implemented within 29 months.

FLORIDA was handed over to the state in June 1992. By the time the system came on-line, it had changed in at least two key aspects from the original design. First, the system was now running on a single mainframe system instead of three. Second, whereas the original design had called for eligibility rulings to be generated overnight, the final system allowed instant, on-line eligibility determinations.

From the time it was first handed over to the state, FLORIDA was a disaster. Response time often approached eight minutes. System crashes, lasting as long as eight hours, were common. Adding to the problems was the fact that, from July 1990 to July 1993,

Florida's food stamps, AFDC, and Medicaid enrollments doubled, from 800,000 to 1.6 million people. FLORIDA was pushed to the limits of its capacity. IBM was commissioned to bring in a new, powerful mainframe system for FLORIDA to run on. As of September 1993, over $173 million had been spent on FLORIDA's development and implementation, almost double the original estimate, and the system was still not complete. Consultants estimated the system had only two more years left to operate, given FLORIDA's then current capacity and projections for the state's welfare enrollments.

The human and financial toll from FLORIDA has easily surpassed those that resulted from New Jersey's vehicle registration system. FLORIDA's use resulted in at least $263 million in welfare over- and underpayments. In 1992, EDS sued HRS for $46 million in payments EDS says it never received. Throughout 1992 and 1993, several key information systems personnel in HRS were fired or forced to resign. In the spring of 1993, the Governor demoted the secretary of HRS and sent the lieutenant governor to HRS to manage the FLORIDA disaster. In August, 1993, a grand jury indicted the former project leader and former deputy secretary of HRS for information systems, for filing false status reports and for falsifying payment reports. In the spring of 1994, all charges were dropped against the former deputy secretary of HRS. The FLORIDA system's former project manager went to trial in May, 1994, and was convicted of two misdemeanor counts of making false statements. The judge in the case overthrew the verdict in June of that year.

In January 1995, the state announced that FLORIDA had been considerably improved, with the error rate for issuing benefits cut to only 12.5%. Yet in October of the same year, the state sued EDS for $60 million for improper business practices in its development of FLORIDA. The state also wanted to ban EDS from ever doing business again in Florida. Two months later, a judge threw out the state's suit. In March 1996, EDS won its 1992 suit against the state. In May of that year, the State of Florida agreed to pay EDS $42.8 million in back payments.

(Compiled from 1993–1996 stories in the *Tallahassee Democrat* and *Information Week*.)

Management support and user involvement are important to implementation success, but they may be overrated compared to other factors that are also important. Research has shown that the link between user involvement and implementation success is sometimes weak (Ives and Olson, 1984). User involvement can help reduce the risk of failure when the system is complex, but user participation in the development process only makes failure more likely when there are financial and time constraints in the development process (Tait and Vessey, 1988). Information systems implementation failures are too common, and the implementation process is too complicated, for the conventional wisdom to be completely correct. The search for better explanations for implementation success and failure has led to two alternative approaches: factor models and political models.

■ FACTOR MODELS OF IMPLEMENTATION SUCCESS

Several research studies have found other factors that are important to a successful implementation process. Ginzberg (1981b) found three additional important factors: commitment to the project, commitment to change, and extent of project definition and planning. Commitment to the project involves managing the system development project so that the problem being solved is well understood and the system being developed to deal with the problem actually solves it. Commitment to change involves being willing to change behaviors, procedures, and other aspects of the organization. The extent of project definition and planning is a measure of how well planned the project is. The more extensive the planning effort, the less likely is implementation failure. In other research, Ginzberg (1981a) uncovered another important factor related to implementation success: user expectations. The more realistic a user's early expectations about a new system and its capabilities, the more likely it is that the user will be satisfied with the new system and actually use it.

Although there are many ways to determine if an implementation has been successful, the two most common and trusted are the extent to which the system is used and the user's satisfaction with the system (Lucas, 1997). Lucas, who has extensively studied information systems implementation, identified six factors that influence the extent to which a system is used (1997):

1. *User's personal stake.* How important the domain of the system is for the user; in other words, how relevant the system is to the work the user performs. The user's personal stake in the system is itself influenced by the level of support management provides for implementation, and by the urgency to the user of the problem addressed by the system. The higher the level of management support and the more urgent the problem, the higher the user's personal stake in the system.

2. *System characteristics.* Includes such aspects of the system's design as ease of use, reliability, and relevance to the task the system supports.

3. *User demographics.* Characteristics of the user, such as age and degree of computer experience.

4. *Organization support.* These are the same issues of support you read about earlier in the section. The better the system support infrastructure, the more likely an individual will be to use the system.

5. *Performance.* What individuals can do with a system to support their work will have an impact on extent of system use. The more users can do with a system and the more creative ways they can develop to benefit from the system the more they will use it. The relationship between performance and use goes both ways. The higher the levels of performance, the more use. The more use, the greater the performance.

6. *Satisfaction.* Use and satisfaction also represent a two-way relationship. The more satisfied the users are with the system, the more they will use it. The more they use it, the more satisfied they will be.

The factors identified by Lucas and the relationships they have to each other are shown in the model in Figure 4-9. In the model, it is easier to see the relationships among the various factors, such as how management support and problem urgency affect the user's personal stake in the system. Notice also that the arrows that show the relationships between use and performance and satisfaction have two heads, illustrating the two-way relationships use has with these factors.

It should be clear that, as an analyst and as someone responsible for the successful implementation of an information system, there are some factors you have more control over than others. For example, you will have considerable influence over the system's characteristics, and you may have some influence over the levels of support that will be provided for users of the system. You have no direct control over a user's demographics, personal stake in the system, management support, or the urgency of the problem to the user. This doesn't

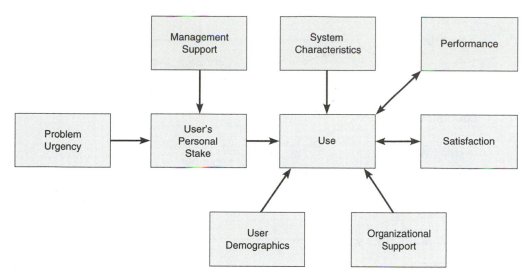

FIGURE 4-9   *Lucas's Model of Implementation Success, 1997 (Adapted with the permission of the McGraw-Hill Companies. All rights reserved.)*

mean you can ignore factors you can't change. On the contrary, you need to understand these factors very well, because you will have to balance them with the factors you *can* change in your system design and in your implementation strategy. You may not be able to change a user's demographics or personal stake in a system, but you can help design the system and your implementation strategy with these factors in mind.

◼ POLITICAL IMPLEMENTATION MODELS

Factor models of the implementation process have helped analysts better understand implementation and why it may or may not be successful. Certainly, taking the different factors into account and realizing how they work together to influence implementation is important. But just as the conventional wisdom about implementation could not explain the whole story, neither can factor models hope to completely explain the implementation process. Political models have been proposed as another perspective to help you understand how a systems development project can succeed. We will use two examples from the implementation literature in MIS to illustrate the usefulness of political models.

Political models assume that individuals who work in an organization have their own self-interested goals, which they pursue in addition to the goals of their departments and the goals of their organizations. Political models also recognize that power is not distributed evenly in organizations. Some workers have more power than others. People may act to increase their own power relative to that of their coworkers and, at other times, people will act to prevent coworkers with more power (such as bosses) from using that power or from gaining more.

Markus (1981) tells the story of a division of an organization where implementation appears to have succeeded. Workers in two manufacturing plants, called Athens and Capital City, were using the new work-in-progress (WIP) system, which made it possible for management to use a planning and forecasting system based in part on the output of the WIP system. Workers from both plants had been involved in the systems development process, especially workers from the Capital City plant. When workers at Athens resisted using the new WIP system, extensive management pressure seemed to force people at Athens to give in and begin using the new system. It appeared that the conventional wisdom about systems development had prevailed: workers had participated in development, and management had been forcefully supportive. Implementation was a success.

Markus presents a political interpretation of the story that provides another explanation of events (see Figure 4-10). She begins by examining the history and power relationships

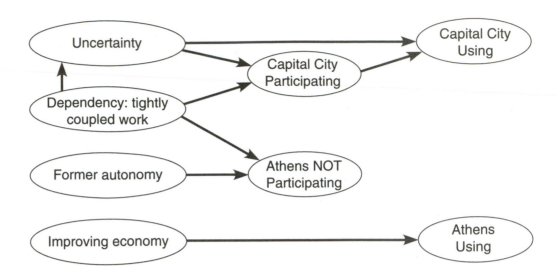

FIGURE 4-10 *Illustrating the Political Explanation of System Success at Athens and Capital City*

within the division and the two plants. Although Athens was the plant that resisted using the new WIP system, they actually had superior information systems support at the beginning of the WIP system development process. Athens had once been a separate company and, until the division head decided new WIP and new planning systems were needed, Athens had been allowed a large amount of autonomy in how it operated. Further, the work performed at the two plants was tightly coupled. The Athens plant manufactured airplane parts, which were then refined and finished at Capital City. Athens' manufacturing process was unpredictable and unreliable and resulted in a high scrap rate. Capital City never knew where Athens was in the manufacturing process for any particular part, and this uncertainty complicated Capital City's efforts to complete its work and finish parts in time for promised delivery dates.

According to Markus, when the opportunity arose to develop a system that would lessen the dependency of Capital City on Athens, people at Capital City were anxious to participate in the system's development. With a new WIP system containing data on Athens' production process, data which Capital City could access, the Capital City plant could greatly reduce the uncertainty associated with its own work. Understandably, the people at Athens were not so enthusiastic, and resisted participating in the development process and using the new system. After the WIP system was completed and installed in both plants, the people at Athens continued to use their old WIP system, much to the chagrin of division management and Capital City. Division management applied a great deal of pressure, until finally Athens began to use the system. Markus states, however, that the people at Athens began using the new system not because of management pressure but because the economy began to improve, and managing the work at Athens became too difficult without the new system.

One lesson to be learned from Markus's case study is that factor-based analyses can only go so far in explaining successful implementation in organizations. History and power relationships are also important considerations. Sometimes political interpretations provide better explanations for the implementation process and why events took the course they did.

Once an information system has been successfully implemented, the importance of documentation grows. A successfully implemented system becomes part of the daily work lives of an organization's employees. Many of those employees will use the system, but others will maintain it and keep it running. Documentation is important for both groups of people.

SUPPORT

As the twentieth century draws to a close, software development and maintenance operations have become major corporate cost centers. In addition, the magnitude and replacement value of software portfolios have become significant tax liabilities. Finally, the on-rushing "Year 2000" problem brings on a set of new and exceptionally large software costs that will have a major financial impact on corporations and government agencies. (Capers Jones, 1997)

In this final section, we discuss systems maintenance, the largest systems development expenditure for many organizations. In fact, more programmers today work on maintenance activities than work on new development. This disproportionate distribution of maintenance programmers is interesting since software does not wear out in a physical manner as do buildings and machines.

There is no single reason why software is maintained; however, most reasons relate to a desire to evolve system functionality in order to overcome internal processing errors or to better support changing business needs. Thus, maintaining a system is a fact of life for most systems. This means that maintenance can begin soon after the system is installed. As with the initial design of a system, maintenance activities are not limited only to software changes but include changes to hardware and business procedures. A question many people have about maintenance relates to how long organizations should maintain a system. Five years? Ten years? Longer? There is no simple answer to this question, but it is most often an issue of economics. In other words, at what point does it make financial sense to discontinue evolving an older system and build or purchase a new one? The focus of a great deal of upper IS management attention is devoted to assessing the trade-offs between maintenance and new development. In this section, we will provide you with a better understanding of the maintenance process and describe the types of issues that must be considered when maintaining systems.

In this section, we also briefly describe the systems maintenance process and the deliverables and outcomes from this process. This is followed by a detailed discussion contrasting the types of maintenance, an overview of critical management issues, and a description of the role of CASE and automated development tools in the maintenance process.

MAINTAINING INFORMATION SYSTEMS

Once an information system is installed, the system is essentially in the maintenance phase of the SDLC. When a system is in the maintenance phase, some person within the systems development group is responsible for collecting maintenance requests from system users and other interested parties, like system auditors, data center and network management staff, and data analysts. Once collected, each request is analyzed to better understand how it will alter the system and what business benefits and necessities will result from such a change. If the change request is approved, a system change is designed and then implemented. As with the initial development of the system, implemented changes are formally reviewed and tested before installation into operational systems.

■ THE PROCESS OF MAINTAINING INFORMATION SYSTEMS

We have drawn the systems development life cycle as the waterfall model where one phase leads to the next with overlap and feedback loops. As shown in Figure 5-1, the maintenance phase is the last phase of the SDLC. Yet, a life cycle, by definition, is circular in that the last activity leads back to the first. This means that the process of maintaining an information system is the process of returning to the beginning of the SDLC (see Figure 5-2) and repeating development steps until the change is implemented.

As shown in Figure 5-1, four major activities occur within maintenance:

1. Obtaining Maintenance Requests
2. Transforming Requests into Changes
3. Designing Changes
4. Implementing Changes

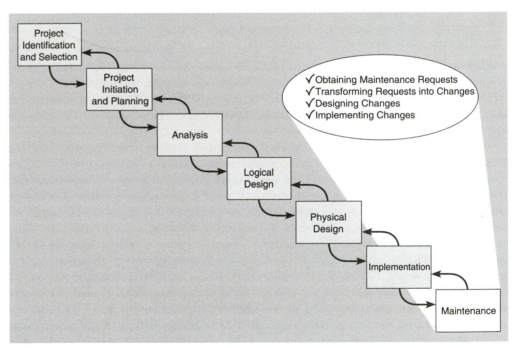

FIGURE 5-1   *Systems Development Life Cycle with Maintenance Phase Highlighted*

Obtaining maintenance requests requires that a formal process be established whereby users can submit system change requests. A user request document called a Systems Service Request (SSR) is shown in Figure 5-3. Most companies have some sort of document like an SSR to request new development, to report problems, or to request new system features with an existing system. When developing the procedures for obtaining maintenance requests, organizations must also specify an individual within the organization to collect these requests and manage their dispersal to maintenance personnel. The process of collecting and dispersing maintenance requests is described in much greater detail later in the section.

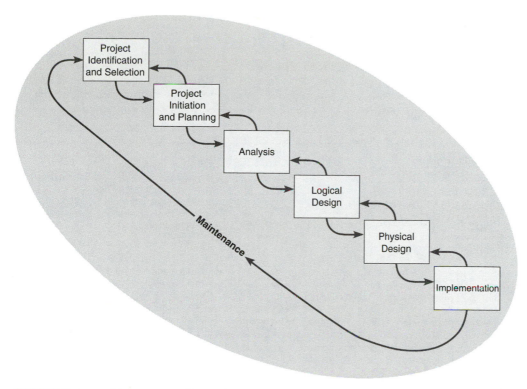

FIGURE 5-2   *Maintenance Phase Makes the System Development Process a Life Cycle*

Once a request is received, analysis must be conducted to gain an understanding of the scope of the request. It must be determined how the request will affect the current system and the duration of such a project. As with the initial development of a system, the size of a maintenance request can be analyzed for risk and feasibility. Next, a change request can be transformed into a formal design change which can then be fed into the maintenance implementation phase. Thus, many similarities exist between the SDLC and the activities within the maintenance process. Figure 5-4 equates SDLC phases to the maintenance activities described previously. The figure shows that the first phase of the SDLC—project identification and selection—is analogous to the maintenance process of obtaining a maintenance request (step 1). SDLC phases project initiation and planning and analysis are analogous to the maintenance process of transforming requests into a specific system change (step 2). The logical and physical design phases of the SDLC, of course, equate to the designing changes process (step 3). Finally, the SDLC phase implementation equates to step 4, implementing changes. This similarity between the maintenance process and the SDLC is no accident. The concepts and techniques used to initially develop a system are also used to maintain it.

■ DELIVERABLES AND OUTCOMES

Since the maintenance phase of the SDLC is basically a subset of the activities of the entire development process, the deliverables and outcomes from the process are the development of a new version of the software and new versions of all design documents developed or modified during the maintenance process. This means that all documents created or modified during the maintenance effort, including the system itself, represent the deliverables and outcomes of the process. Those programs and documents that did not change may also be part of the new system. Since most organizations archive prior versions of systems, all prior programs and documents must be kept to ensure the proper versioning of the system. This enables prior versions of the system to be recreated if needed. A more detailed discussion of configuration management and change control is presented later in the section.

Pine Valley Furniture
System Service Request

REQUESTED BY   Juanita Lopez   DATE   November 1, 1998

DEPARTMENT   Purchasing, Manufacturing Support

LOCATION   Headquarters, 1-322

CONTACT   Tel: 4-3267   FAX: 4-3270   e-mail: jlopez

TYPE OF REQUEST   URGENCY

[X]   New System   [   ]   Immediate – Operations are impaired or opportunity lost

[   ]   System Enhancement   [   ]   Problems exist, but can be worked around

[   ]   System Error Correction   [X]   Business losses can be tolerated until new system installed

PROBLEM STATEMENT

Sales growth at PVF has caused greater volume of work for the manufacturing support unit within Purchasing. Further, more concentration on customer service has reduced manufacturing lead times, which puts more pressure on purchasing activities. In addition, cost-cutting measures force Purchasing to be more agressive in negotiating terms with vendors, improving delivery times, and lowering our investments in inventory. The current modest systems support for manufacturing purchasing is not responsive to these new business conditions. Data are not available, information cannot be summarized, supplier orders cannot be adequately tracked, and commodity buying is not well supported. PVF is spending too much on raw materials and not being responsive to manufacturing needs.

SERVICE REQUEST

I request a thorough analysis of our current operations with the intent to design and build a completely new information system. This system should handle all purchasing transactions, support display and reporting of critical purchasing data, and assist purchasing agents in commodity buying.

IS LIAISON   Chris Martin (Tel: 4-6204   FAX: 4-6200   e-mail: cmartin)

SPONSOR   Sal Divario, Director, Purchasing

------------------------TO BE COMPLETED BY SYSTEMS PRIORITY BOARD----------------------

[   ]   Request approved   Assigned to _____

                                 Start date _____

[   ]   Recommend revision

[   ]   Suggest user development

[   ]   Reject for reason _____

FIGURE 5-3   *Systems Request for Purchasing Fulfillment System (Pine Valley Furniture)*

     Because of the similarities between the steps, deliverables, and outcomes of new development and maintenance, you may be wondering how to distinguish between these two processes. One difference is that maintenance reuses most existing system modules in producing the new system version. Other distinctions are that we develop a new system when there is a change in the hardware or software platform or when fundamental assumptions and properties of the data, logic, or process models change.

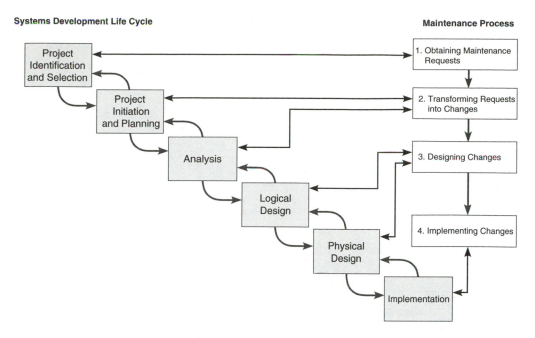

Systems Development Life Cycle

Maintenance Process

FIGURE 5-4 *Maintenance Activities in Relation to the SDLC*

CONDUCTING SYSTEMS MAINTENANCE

A significant portion of the expenditures for information systems within organizations does not go to the development of new systems but to the maintenance of existing systems. We will describe various types of maintenance, factors influencing the complexity and cost of maintenance, alternatives for managing maintenance, and the role of CASE during maintenance. Given that maintenance activities consume the majority of information systems-related expenditures, gaining an understanding of these topics will yield numerous benefits to your career as an information systems professional.

■ TYPES OF MAINTENANCE

There are several types of maintenance that you can perform on an information system (see Table 5.1). By maintenance, we mean the fixing or enhancing of an information system. Corrective maintenance refers to changes made to repair defects in the design, coding, or implementation of the system. For example, if you had recently purchased a new home, corrective maintenance would involve repairs made to things that had never worked as designed, such as a faulty electrical outlet or misaligned door. Most corrective maintenance problems surface soon after installation. When corrective maintenance problems surface, they are typically urgent and need to be resolved to curtail possible interruptions in normal business

TABLE 5.1
Types of Maintenance

Type	Description
Corrective	Repair design and programming errors
Adaptive	Modify system to environmental changes
Perfective	Evolve system to solve new problems or take advantage of new opportunities
Preventive	Safeguard system from future problems

activities. Of all types of maintenance, corrective accounts for as much as 75 percent of all maintenance activity (Andrews and Leventhal, 1993). This is unfortunate because corrective maintenance adds little or no value to the organization; it simply focuses on removing defects from an existing system without adding new functionality (see Figure 5-5).

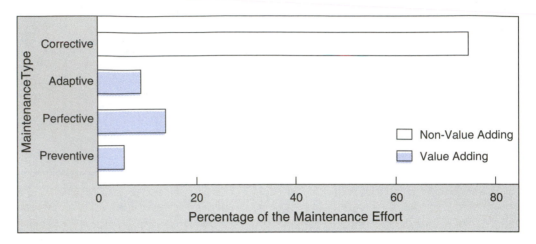

FIGURE 5-5　*Types of Maintenance (Adapted from Andrews and Leventhal, 1993)*

Adaptive maintenance involves making changes to an information system to evolve its functionality to changing business needs or to migrate it to a different operating environment. Within a home, adaptive maintenance might be adding storm windows to improve the cooling performance of an air conditioner. Adaptive maintenance is usually less urgent than corrective maintenance because business and technical changes typically occur over some period of time. Contrary to corrective maintenance, adaptive maintenance is generally a small part of an organization's maintenance effort but does add value to the organization.

Perfective maintenance involves making enhancements to improve processing performance, interface usability, or to add desired, but not necessarily required, system features ("bells and whistles"). In our home example, perfective maintenance would be adding a new room. Many system professionals feel that perfective maintenance is not really maintenance but new development.

Preventive maintenance involves changes made to a system to reduce the chance of future system failure. An example of preventive maintenance might be to increase the number of records that a system can process far beyond what is currently needed or to generalize how a system sends report information to a printer so that the system can easily adapt to changes in printer technology. In our home example, preventive maintenance could be painting the exterior to better protect the home from severe weather conditions. As with adaptive maintenance, both perfective and preventive maintenance are typically a much lower priority than corrective maintenance. Over the life of a system, corrective maintenance is most likely to occur after initial system installation or after major system changes. This means that adaptive, perfective, and preventive maintenance activities can lead to corrective maintenance activities if not carefully designed and implemented.

■ THE COST OF MAINTENANCE

Information systems maintenance costs are a significant expenditure. For some organizations, as much as 80 percent of their information systems budget is allocated to maintenance activities (Pressman, 1992). Additionally, the proportion of systems expenditures has been rising due to the fact that many organizations have accumulated more and more older systems that require more and more maintenance.

For example, Figure 5-6 shows that in the 1970s, most of an organization's information systems' expenditures were allocated to new development rather than to maintenance.

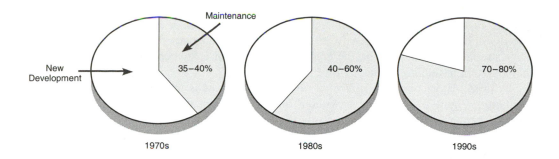

FIGURE 5-6   *New Development Versus Maintenance as a Percent of Software Budget (Adapted from Pressman, 1987)*

However, over the years, this mix has changed so that the majority of expenditures are now earmarked for maintenance. This means that you must understand the factors influencing the maintainability of systems. Maintainability is the ease with which software can be understood, corrected, adapted, and enhanced. Systems with low maintainability result in uncontrollable maintenance expenses.

Numerous factors influence the maintainability of a system. These factors, or cost elements, will determine the extent to which a system has high or low maintainability. Table 5.2 shows numerous elements that influence the cost of maintenance, many of which you can influence as a systems analyst. Of these factors, three are most significant: number of latent defects, number of customers, and documentation quality.

TABLE 5.2
Cost Elements of Maintenance

Element	Description
Defects	Number of unknown defects in a system when it is installed
Customers	Number of different customers that a maintenance group must support
Documentation	Quality of technical system documentation including test cases
Personnel	Number and quality of personnel dedicated to the support and maintenance of a system
Tools	Software development tools, debuggers, hardware, and other resources
Software Structure	Structure and maintainability of the software

Adapted from Jones, 1986

The number of latent defects refers to the number of unknown errors existing in the system after it is installed. Because corrective maintenance accounts for most maintenance activity, the number of latent defects in a system influences most of the costs associated with maintaining a system. If there are no errors in the system after it is installed, then maintenance costs will be relatively low. If there are a large number of defects in the system when it is installed, maintenance costs will likely be high.

A second factor influencing maintenance costs is the number of customers for a given system. In general, the greater the number of customers, the greater the maintenance costs. For example, if a system has only one customer, problem and change requests will come from only one source. A single customer also makes it much easier for the maintenance group to know how the system is actually being used and the extent to which users are adequately trained. If the system fails, the maintenance group is only notifying, supporting, and retraining a small group of people and updating their programs and documentation. On the other hand, for a system with thousands of users, change requests (possibly contradictory or incompatible) and error reports come from many places, and notification of problems, customer support, and system and documentation updating become a much more significant problem. For example, notifying all customers of a problem is fast and easy when you have a single customer whom you can contact using a telephone, fax, or electronic mail message. Yet, it will be difficult and expensive to quickly and easily contact thousands of users about a catastrophic problem. In sum, the greater the number of customers, the more critical it is that a system have high maintainability.

A third major contributing factor to maintenance costs is the quality of system documentation. Figure 5-7 shows that without quality documentation, maintenance effort can increase exponentially. In conclusion, numerous factors will influence the maintainability and thus the overall costs of system maintenance. System professionals have found that the number of defects in the installed system drive all other cost factors. This means that it is important that you remove as many errors as possible before installation.

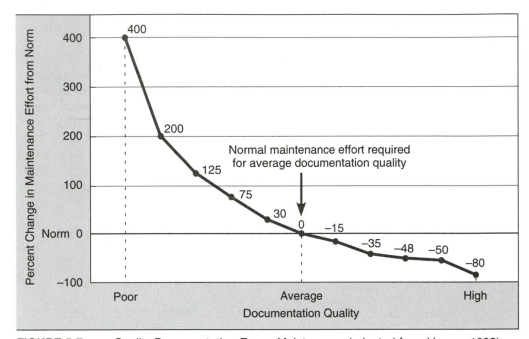

FIGURE 5-7   *Quality Documentation Eases Maintenance (adapted from Hanna, 1992)*

The quality of the maintenance personnel also contributes to the cost of maintenance. In some organizations, the best programmers are assigned to maintenance. Highly skilled programmers are needed because the maintenance programmer is typically not the original programmer and must quickly understand and carefully change the software. Tools, such as those that can automatically produce system documentation where none exists, can also lower maintenance costs. Finally, well-structured programs make it much easier to understand and fix programs.

■ MANAGING MAINTENANCE

As maintenance activities consume more and more of the systems development budget, maintenance management has become increasingly important. For example, Table 5.3 shows the worldwide growth of programmers who are working on new development versus maintenance. These data show that the number of people working on maintenance has now surpassed the number working on new development. Maintenance has become the largest segment of programming personnel, and this implies the need for careful management. We will address this concern by discussing several topics related to the effective management of systems maintenance.

TABLE 5.3
Worldwide Totals of Programmers Working on New Development versus Maintenance

Year	Programmers on New Programs	Programmers on Maintenance
1950	90	10
1960	8,500	1,500
1970	65,000	35,000
1980	1,200,000	800,000
1990	3,000,000	4,000,000
2000	4,000,000	6,000,000

Adapted from Jones, 1986

MANAGING MAINTENANCE PERSONNEL   One concern with managing maintenance relates to personnel management. Historically, many organizations had a "maintenance group" that was separate from the "development group." With the increased number of maintenance personnel, the development of formal methodologies and tools, changing organizational forms, end-user computing, and the widespread use of very high-level languages for the development of some systems, organizations have rethought the organization of maintenance and development personnel (Chapin, 1987). In other words, should the maintenance group be separated from the development group? Or, should the same people who build the system also maintain it? A third option is to let the primary end-users of the system in the functional units of the business have their own maintenance personnel. The advantages and disadvantages to each of these organizational structures are summarized in Table 5.4.

In addition to the advantages and disadvantages listed in Table 5.4, there are numerous other reasons why organizations should be concerned with how they manage and assign maintenance personnel. One key issue is that many systems professionals don't want to perform maintenance because they feel that it is more exciting to build something new rather than change an existing system (Martin, Brown, DeHayes, Hoffer, and Perkins, 1999). In other words, maintenance work is often viewed as "cleaning up someone else's mess." Also, organizations have historically provided greater rewards and job opportunities to those performing new development, thus making people shy away from maintenance-type careers. As a result, no matter how an organization chooses to manage its maintenance group—separate, combined, or functional—it is now common to rotate individuals in and out of maintenance activities. This rotation is believed to lessen the negative feelings about maintenance work and to give personnel a greater understanding of the difficulties and relationships between new development and maintenance.

TABLE 5.4
Advantages and Disadvantages of Different Maintenance Organizational Structures

Type	Advantages	Disadvantages
Separate	Formal transfer of systems between groups improves the system and documentation quality	All things cannot be documented, so the maintenance group may not know critical information about the system
Combined	Maintenance group knows or has access to all assumptions and decisions behind the system's original design	Documentation and testing thoroughness may suffer due to a lack of a formal transfer of responsibility
Functional	Personnel have a vested interest in effectively maintaining the system and have a better understanding of functional requirements	Personnel may have limited job mobility and lack access to adequate human and technical resources

MEASURING MAINTENANCE EFFECTIVENESS   A second management issue is the measurement of maintenance effectiveness. As with the effective management of personnel, the measurement of maintenance activities is fundamental to understanding the quality of the development and maintenance efforts. To measure effectiveness, you must measure these factors:

- Number of failures
- Time between each failure
- Type of failure

Measuring the number and time between failures will provide you with the basis to calculate a widely used measure of system quality. This metric is referred to as the mean time between failures (MTBF). As its name implies, the MTBF measure shows the average length of time between the identification of one system failure until the next. Over time, you should expect the MTBF value to rapidly increase after a few months of use (and corrective maintenance) of the system (see Figure 5-8 for an example of the relationship between MTBF and age of a system). If the MTBF does not rapidly increase over time, it will be a signal to management that major problems exist within the system that are not being adequately resolved through the maintenance process.

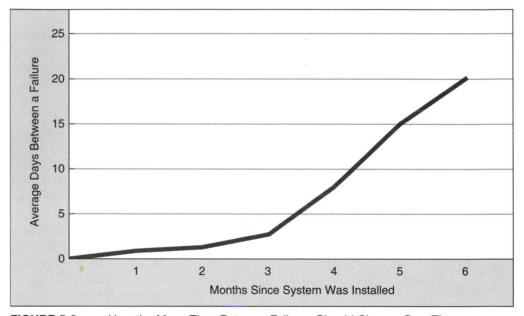

FIGURE 5-8   *How the Mean Time Between Failures Should Change Over Time*

A more revealing method of measurement is to examine the failures that are occurring. Over time, logging the types of failures will provide a very clear picture of where, when, and how failures occur. For example, knowing that a system repeatedly fails logging new account information to the database when a particular customer is using the system can provide invaluable information to the maintenance personnel. Were the users adequately trained? Is there something unique about this user? Is there something unique about an installation that is causing the failure? What activities were being performed when the system failed?

Tracking the types of failures also provides important management information for future projects. For example, if a higher frequency of errors occurs when a particular development environment is used, such information can help guide personnel assignments, training courses, or the avoidance of a particular package, language, or environment during future development. The primary lesson here is that without measuring and tracking maintenance activities, you cannot gain the knowledge to improve or know how well you are doing relative to the past. To effectively manage and to continuously improve, you must measure and assess performance over time.

CONTROLLING MAINTENANCE REQUESTS   Another maintenance activity is managing maintenance requests. There are various types of maintenance requests—some correct minor or severe defects in the system while others improve or extend system functionality. From a management perspective, a key issue is deciding which requests to perform and which to ignore. Since some requests will be more critical than others, some method of prioritizing requests must be determined.

Figure 5-9 shows a flow chart that suggests one possible method you could apply for dealing with maintenance change requests. First, you must determine the type of request. If, for example, the request is an error—that is, a corrective maintenance request—then the flow chart shows that a question related to the error's severity must be asked. If the error is "very severe," then the request has top priority and is placed at the top of a queue of tasks waiting to be performed on the system. In other words, for an error of high severity, repairs to remove it must be made as soon as possible. If, however, the error is considered "nonsevere," then the change request can be categorized and prioritized based upon its type and relative importance.

If the change request is not an error, then you must determine whether the request is to adapt the system to technology changes and/or business requirements or to enhance the system so that it will provide new business functionality. For adaptation requests, they too will need to be evaluated, categorized, prioritized, and placed in the queue. For enhancement-type requests, they must first be evaluated to see whether they are aligned with future business and information systems' plans. If not, the request will be rejected and the requester will be informed. If the enhancement appears to be aligned with business and information systems plans, it can then be prioritized and placed into the queue of future tasks. Part of the prioritization process includes estimating the scope and feasibility of the change. Techniques used for assessing the scope and feasibility of entire projects should be used when assessing maintenance requests.

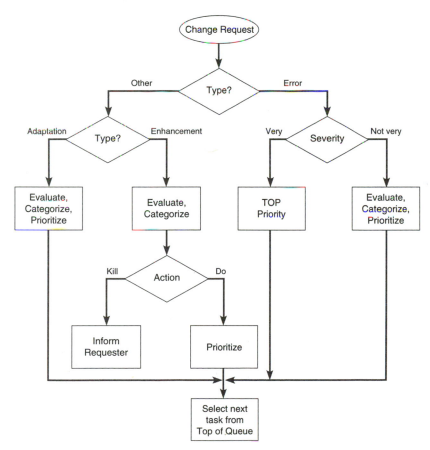

FIGURE 5-9   *Flow Chart of How to Control Maintenance Requests (Adapted from Pressman, 1992)*

The queue of maintenance tasks for a given system is dynamic—growing and shrinking based upon business changes and errors. In fact, some lower-priority change requests may never be accomplished, since only a limited number of changes can be accomplished at a given time. In other words, changes in business needs between the time the request was made and when the task finally rises to the top of the queue may result in the request being deemed unnecessary or no longer important given current business directions. Thus, managing the queue of pending tasks is an important activity.

To better understand the flow of a change request, see Figure 5-10. Initially, an organizational group that uses the system will make a request to change the system. This request flows to the project manager of the system (labeled 1). The project manager evaluates the request in relation to the existing system and pending changes, and forwards the results of this evaluation to the System Priority Board (labeled 2). This evaluation will also include a feasibility analysis that includes estimates of project scope, resource requirements, risks, and other relevant factors. The board evaluates, categorizes, and prioritizes the request in relation to both the strategic and information systems plans of the organization (labeled 3). If the board decides to kill the request, the project manager informs the requester and explains the rationale for the decision (labeled 4). If the request is accepted, it is placed in the queue of pending tasks. The project manager then assigns tasks to maintenance personnel based upon their availability and task priority (labeled 5). On a periodic basis, the project manager prepares a report of all pending tasks in the change request queue. This report is forwarded to the priority board where they reevaluate the requests in light of the current business conditions. This process may result in removing some requests or reprioritizing others.

Although each change request goes through an approval process as depicted in Figure 5-10, changes are usually implemented in batches, forming a new release of the software. It is too difficult to manage a lot of small changes. Further, batching changes can reduce maintenance work when several change requests affect the same or highly related modules. Frequent releases of new system versions may also confuse users if the appearance of displays, reports, or data entry screens changes.

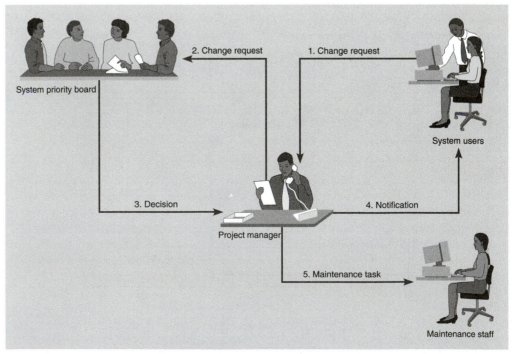

FIGURE 5-10   *How a Maintenance Request Moves Through an Organization (Adapted from Pressman, 1987)*

CONFIGURATION MANAGEMENT   A final aspect of managing maintenance is configuration management, which is the process of assuring that only authorized changes are made to a system. Once a system has been implemented and installed, the programming code used to construct the system represents the baseline modules of the system. The baseline modules are the software modules for the most recent version of a system where each module has passed the organization's quality assurance process and documentation standards. A system librarian controls the baseline source code modules. If maintenance personnel are assigned to make changes to a system, they must first check out a copy of the baseline system modules because no one is allowed to directly modify the baseline modules. Only those that have been checked out and have gone through a formal check-in process can reside in the library. Before any code can be checked back in to the librarian, the code must pass the quality control procedures, testing, and documentation standards established by the organization.

When various maintenance personnel working on different maintenance tasks complete each task, the librarian notifies those still working that updates have been made to the baseline modules. This means that all tasks being worked on must now incorporate the latest baseline modules before being approved for check-in. Following a formal process of checking modules out and in, a system librarian helps to assure that only tested and approved modules become part of the baseline system. It is also the responsibility of the librarian to keep copies of all prior versions of all system modules, including the build routines needed to construct *any version* of the system that *ever* existed. It may be important to reconstruct old versions of the system if new ones fail, or to support users that cannot run newer versions on their computer system.

Special software systems have been created to manage system configuration and version control activities (see box "Configuration Management Tools"). This software is increasingly necessary as the change control process is complicated in organizations deploying several different networks, operating systems, languages, and database management systems in which there may be many concurrent versions of an application, each for a different platform. One function of this software is to control access to libraried modules. Each time a module is checked out or in, this activity is recorded, after being authorized by the librarian. The software helps the librarian track that all necessary steps have been followed before a new module is released to production, including all integration tests, documentation updates, and approvals.

Configuration Management Tools

There are two general kinds of configuration management tools: revision control and source code control. With revision control tools, each system module file is "frozen" (unchangeable) to a specific version level—or "floating"—a programmer may check out, lock, and modify. Only the most recent version of a module is stored; previous versions are reconstructed when needed by applying changes in reverse order. Source code control tools extend the above description to address inter-related files. These tools also help in rebuilding any historic version of a system by recompiling the proper source code modules. They also trace an executable code module back to its source code version.

ROLE OF CASE AND AUTOMATED DEVELOPMENT TOOLS IN MAINTENANCE

In traditional systems development, much of the time is spent on coding and testing. When software changes are approved, code is first changed and then tested. Once the functionality of the code is assured, the documentation and specification documents are updated to reflect system changes. Over time, the process of keeping all system documentation "current" can be a very boring and time-consuming activity that is often neglected. This neglect makes future maintenance by the same or *different* programmers difficult at best.

A primary objective of using CASE and other automated tools for systems development and maintenance is to radically change the way in which code and documentation are modified and updated. When using an integrated development environment, analysts maintain design documents such as data flow diagrams and screen designs, not source code. In other words, design documents are modified and then code generators automatically create a new version of the system from these updated designs. Also, since the changes are made at the design specification level, most documentation changes such as an updated data flow diagram will have already been completed during the maintenance process itself. Thus, one of the biggest advantages to using CASE, for example, is its benefits during system maintenance.

In addition to using general automated tools for maintenance, two special-purpose tools, reverse engineering and reengineering tools, are primarily used to maintain older systems that have incomplete documentation or that were developed prior to CASE use. These tools are often referred to as *design recovery tools* since their primary benefit is to create high-level design documents of a program by reading and analyzing its source code.

Recall that reverse engineering tools are those that can create a representation of a system or program module at a design level of abstraction. For example, reverse engineering tools read program source code as input, perform an analysis, and extract information such as program control structures, data structures, and data flow. Once a program is represented at a design level using both graphical and textual representations, the design can be more effectively restructured to current business needs or programming practices by an analyst. Similarly, reengineering tools extend reverse engineering tools by automatically, or interactively with a systems analyst, altering an existing system in an effort to improve its quality or performance.

TRAINING AND SUPPORTING USERS

Training and support are two aspects of an organization's computing infrastructure (Kling and Iacono, 1989). Computing infrastructure is made up of all the resources and practices required to help people adequately use computer systems to do their primary work (Kling and Scacchi, 1982; Gasser, 1986). It is analogous to the infrastructure of water mains, electric power lines, streets, and bridges that form the foundation for providing essential services in a city. Henderson and Treacy (1986) identify infrastructure as one of four fundamental issues IS managers must address. They suggest that training and support are most important in the early stages of end-user computing growth and less so later on. Rockart and Short (1989) cite the "development and implementation of a general, and eventually 'seamless,' information technology infrastructure" as a major demand on information technology. They list the creation of an effective information technology infrastructure as one of the five key issues for senior organizational managers in the 1990s. Thus, training and support are critical for the success of an information system. As the person whom the user holds responsible for the new system, you and other analysts on the project team must ensure that high-quality training and support are available.

Although training and support can be talked about as if they are two separate things, in organizational practice the distinction between the two is not all that clear, as the two sometimes overlap. After all, both deal with learning about computing. It is clear that support mechanisms are also a good way to provide training, especially for intermittent users of a system (Eason, 1988). Intermittent or occasional system users are not interested in, nor would they profit from, typical user training methods. Intermittent users must be provided with "point of need support," specific answers to specific questions at the time the answers are needed. A variety of mechanisms, such as the system interface itself and on-line help facilities, can be designed to provide both training and support at the same time.

The value of support is often underestimated. Few organizations invest heavily in support staff, which can lead to users solving problems for themselves or somehow working around them (Gasser, 1986). Adequate user support may be essential for successful information system development, however. One study found that user satisfaction with support provided by the information systems department was the factor most closely related to overall satisfaction with user development of computer-based applications (Rivard and Huff, 1988).

■ TRAINING INFORMATION SYSTEM USERS

Computer use requires skills, and training people to use computer applications can be expensive for organizations (Kling and Iacono, 1989). Training of all types is a major activity in American corporations (they spent $48 billion on training in 1993), but information systems training is often neglected. Many organizations tend to underinvest in computing skills training. It is true that some organizations institutionalize high levels of information system training, but many others offer no systematic training at all.

Some argue that information systems departments are similar to hospitals: both are high-technology environments, both are staffed by well-educated professionals, both are capital-intensive, and both have less than adequate "bedside manners" (Schrage, 1993). In this scenario, users, like patients, are seen as problems to be solved rather than as people. One response to this issue might be to spread more evenly the cost of computer training. Even though users come from all parts of the organization, the information systems department is often stuck paying for most of the computer training itself. Attitudes in information systems towards users and the availability of high-quality training might change if other organizational departments begin to pay their share of the computer training bill.

Others argue that it is difficult to demonstrate any direct benefits from training once the training is complete, since managers are more likely to continually upgrade hardware and software (de Jager, 1994). Progress from upgrading is easy to measure—just look at all the empty cartons—but upgrading may not result in the increases in productivity you would expect. Training users to be effective with the systems they have now may be a more cost-effective way to increase productivity: "If the goal is maximum productivity at the lowest cost, then a day of training usually delivers more productivity per dollar than costly hardware and software upgrades" (de Jager, 1994, p. 86). Though not always taken seriously in practice, the value and cost-effectiveness of information systems training should not be underestimated.

The type of necessary training will vary by type of system and expertise of users. The list of potential topics from which you must determine if training will be useful include the following:

- Use of the system (e.g., how to enter a class registration request)
- General computer concepts (e.g., computer files and how to copy them)
- Information system concepts (e.g., batch processing)
- Organizational concepts (e.g., FIFO inventory accounting)
- System management (e.g., how to request changes to a system)
- System installation (e.g., how to reconcile current and new systems during phased installation)

As you can see from this partial list, there are potentially many topics that go beyond simply how to use the new system. It may be necessary for you to develop training for users in other areas so that users will be ready, conceptually and psychologically, to use the new system. Some training, like concept training, should begin early in the project since this training can assist in the "unfreezing" element of the organizational change process.

Each element of training can be delivered in a variety of ways. Table 5.5 lists the most common training methods used by information system departments a few years ago. Despite the importance and value of training, most of the methods listed in Table 5.5 are under-utilized in many organizations. Users primarily rely on just one of these delivery modes: more often than not, users turn to the resident expert and to fellow users for training, as shown in

TABLE 5.5
Seven Common Methods for Computer Training in 1987

1. Tutorial—one person taught at a time
2. Course—several people taught at a time
3. Computer-aided instruction
4. Interactive training manuals—combination of tutorials and computer-aided instruction
5. Resident expert
6. Software help components
7. External sources, such as vendors

Adapted from Nelson and Cheney, 1987

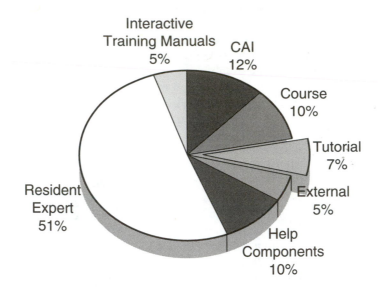

FIGURE 5-11   *Frequency of Use of Computer Training Methods (Nelson & Cheney, 1987)*

how often people use each mode listed in Table 5.5 (see Figure 5-11). One study reported that 89.4 percent of end-users consulted their colleagues about how to use microcomputers, but only 48 percent consulted central information systems staff (Lee, 1986). Users are more likely to turn to local experts for help than to the organization's technical support staff because the local expert understands both the users' primary work and the computer systems they use (Eason, 1988). Given their dependence on fellow users for training, it should not be surprising that end-users describe their most common mode of computer training as "self-training" (Nelson and Cheney, 1987). Self-training has been found to be associated with particular sets of user skills: using application development software, using packaged application software, data communication, using hardware, using operating systems, and graphics skills. The last four sets of skills are also highly associated with company-provided training. There appear to be areas of training best accomplished by centralized, company-provided training on the one hand and by self-training on the other.

One conclusion from the experience with user training methods is that an effective strategy for training on a new system is to first train a few key users and then organize training programs and support mechanisms which involve these users to provide further training, both formal and on demand. Often, training is most effective if you customize it to particular user groups, and the lead trainers from these groups are in the best position to do this.

While individualized training is expensive and time-consuming, technological advances and decreasing costs have made this type of training more feasible. Similarly, the number of training modes used by information systems departments today has increased dramatically beyond what is listed in Table 5.5. Training modes now include videos, interactive television for remote training, multimedia training, on-line tutorials, and electronic performance support systems (EPSS). These may be delivered via videotapes, CD-ROMS, company intranets, and the Internet.

Electronic performance support systems are on-line help systems that go beyond simply providing help—they embed training directly into a software package (Cole et al., 1997). An EPSS may take on one or more forms: they can be an on-line tutorial, provide hypertext-based access to context-sensitive reference material, or consist of an expert system shell that acts as a coach. The main idea behind the development of an EPSS is that the user never has to leave the application to get the benefits of training. Users learn a new system or unfamiliar features at their own pace and on their own machines, without having to lose work time to remote group training sessions. Furthermore, this learning is on-demand when the user is most motivated to learn, since the user has a task to do. EPSS is sometimes referred to as

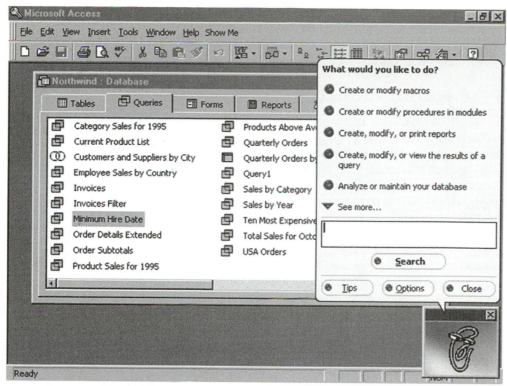

FIGURE 5-12   *A Microsoft Office Assistant™ Note*

"just-in-time knowledge." One example of an EPSS with which you may be familiar is Microsoft's Office Assistant™. Office Assistants (Figure 5-12) are animated characters that come up on top of such applications as Access™ and Word™. You ask questions and the Office Assistants return answers that provide educational information, such as graphics, examples, and procedures, as well as hypertext nodes for jumping to related help topics. Microsoft's Office Assistants communicate with the application you are running to see where you are, so you can determine, by reading the context-sensitive information, if what you want to do is possible from your present location. Some EPSS environments actually walk the user step-by-step through the task, coaching the user on what to do or allowing the user to get additional on-line assistance at any point.

Training for information systems is increasingly being made available both over company intranets and over the Internet. International Data Corporation projects that the on-line training industry will grow to a $28 billion business by the year 2001 (Dillon, 1997). Individual companies may prepare the training and make it available with the help of vendors who convert the content to work on the Internet. This is the case with Coopers & Lybrand's Tax News Network, which is available to 2000 subscribers as well as 3000 company staffers. Coopers & Lybrand features a two-hour training course on new tax legislation on its password-protected site. Alternatively, an organization may prepare its own training content using course authoring software. An example of this approach is Stanford University's Education Program for Gifted Youth, which Stanford provides over the Internet from its own servers. Still a third alternative is to access training provided by third party vendors. An example of this service is provided by CBT Systems, which offers 279 courses over the Internet. CBT Systems specializes in training on information technology products from multiple vendors, including Microsoft, Oracle, Novell, Powersoft, and SAP. Accessing training over the Internet has the potential to save companies thousands of dollars each year in training costs, especially when it comes to information technology training. Instead of having to send personnel off-site for weeks and pay their travel expenses, companies can gain access to Internet training for lower cost and personnel can get the training at their desks.

As both training and support for computing are increasingly able to be delivered on-line in modules, with some embedded in software packages and applications (as is the case for EPSS), the already blurred distinction between them blurs even more. Some of the issues most particular to computer user support are examined in the next section.

■ SUPPORTING INFORMATION SYSTEM USERS

Historically, computing support for users has been provided in one of a few forms: paper, on-line versions of paper-based support, vendors, or by other people who work for the same organization. As we stated earlier, support, whatever its form, has often been inadequate for users' needs. Yet users consider support to be extremely important. A 1993 J. D. Power and Associates survey found user support to be the number one criterion contributing to user satisfaction with personal computing, cited by 26 percent of respondents as the most important factor (Schurr, 1993).

As organizations moved away from mainframe-based computing to increased reliance on personal computers in the 1980s, the need for support among users also increased. More support was made available on-line, but organizations also began to provide institutionalized user support in the form of information centers and help desks. An information center comprises a group of people who can answer questions and assist users with a wide range of computing needs, including the use of particular information systems. Information center staff might do the following:

- Install new hardware or software
- Consult with users writing programs in fourth-generation languages
- Extract data from organizational databases onto personal computers
- Set up user accounts
- Answer basic on-demand questions
- Provide a demonstration site for viewing hardware and software
- Work with users to submit system change requests

When you expect an organizational information center to help support a new system, you will likely want to train the information center staff as soon as possible on the system. Information center staff will need additional training periodically as new system features are introduced, new phases of a system are installed, or system problems and workarounds are identified. Even if new training is not required, information center staff should be aware of system changes, since such changes may result in an increase in demand for information center services.

Personnel in an information center or help desk (we will discuss help desks in a later section in this section) were typically drawn from the ranks of information systems workers or knowledgeable users in functional area departments. There was rarely any type of formal training for people to learn how to work in the support area and, in general, this remains true today. As organizations move to client/server architectures, their needs for support have increased even more than during the introduction of PCs, and organizations find themselves in the position of having to rely more than ever on vendor support (From, E 1993). This increased need for support comes in part from the lack of standards governing client/server products and the resulting need to make equipment and software from different vendors compatible. Vendors are able to provide the necessary support, but as they have shifted their offerings from primarily expensive mainframe packages to inexpensive off-the-shelf software, they find they can no longer bear the cost of providing the support for free. Most vendors now charge for support, and many have instituted 900 numbers or sell customers unlimited support for a given monthly or annual charge.

Automating Support   In an attempt to cut the costs of providing support and to catch up with the demand for additional support services, vendors have automated much of their support offerings. Common methods for automating support include on-line support forums, bulletin board systems, on-demand fax, and voice-response systems (Schneider, 1993). On-line support forums provide users access to information on new releases, bugs, and tips for more effective usage. Forums are offered on on-line services such as America OnLine™, over the Internet, and over company intranets. On-demand fax allows users to order support information through an 800 number and receive that information instantly over their fax machines. Finally, voice-response systems allow users to navigate option menus that lead to prerecorded messages about usage, problems, and workarounds. Organizations have established similar support mechanisms for systems developed or purchased by the organization. Internal e-mail, group support systems, and office automation can be used to support such capabilities within an organization.

Other organizations can follow the lead of vendors in automating some of their support offerings in similar ways. Some software vendors work with large accounts and corporations to set up or enhance their own support offerings. The cost to the corporation can go from $5,000 to as much as $20,000. Offerings include access to knowledge bases about a vendor's products, electronic support services, single point of contact, and priority access to vendor support personnel (Schneider, 1993). Product knowledge bases include all of the technical and support information about vendor products and provide additional information for on-site personnel to use in solving problems. Some vendors allow users to access some of the same information that is available in on-line forums through bulletin boards which the vendors set up and maintain themselves and on Internet sites. Many vendors now supply complete user and technical documentation on CD-ROM, including periodic updates, so that a user organization can provide this library of documentation, bug reports, workaround notices, and notes on undocumented features on-line to all internal users. Electronic support services include all of the vendor support services discussed earlier, but tailored specifically for the corporation. The single point of contact is a system engineer who is often based on site and serves as a liaison between the corporation and the vendor. Finally, priority access means that corporate workers can always get help via telephone or e-mail from a person at the vendor company, usually within a prespecified response time of four hours or less.

Such vendor-enhanced support is especially appropriate in organizations where a wide variety of a particular vendor's products is in use, or where most in-house application development either utilizes vendor products as components of the larger system or where the vendor's products are themselves used as the basis for applications. An example of the former would be the case where an organization has set up a client/server architecture based on a particular vendor's SQL server and application programmer interfaces (APIs). Which applications are developed in-house to run under the client/server architecture depends heavily on the server and APIs, and direct vendor support dealing with problems in these components would be very helpful to the enterprise information systems staff. An example of the second would include order entry and inventory control application systems developed using Microsoft's Access™ or Excel™. In this case, the system developers and users, who are sometimes the same people for such package-based applications, can benefit considerably from directly questioning vendor representatives about their products.

PROVIDING SUPPORT THROUGH A HELP DESK   Whether assisted by vendors or going it alone, the center of support activities for a specific information system in many organizations is the help desk. A help desk is an information systems department function, staffed by IS personnel, possibly part of the information center unit. The help desk is the first place users should call when they need assistance with an information system. The help desk staff either deals with the users' questions or refers the users to the most appropriate person.

For many years, a help desk was the dumping ground for people IS managers did not know what else to do with. Turnover rates were high because the position was sometimes little more than a complaints department, the pay was low, and burnout rates were high. In today's information systems-dependent enterprises, however, this situation has changed. Help desks are gaining new respect as management comes to appreciate the special combination of technical skills and people skills needed to make good help desk staffers. In fact, a recent survey reveals that the top two valued skills for help desk personnel are related to communication and customer service (Crowley, 1993).

Help desk personnel (as well as the personnel in the more general information center) need to be good at communicating with users, by listening to their problems and intelligently communicating potential solutions. These personnel also need to understand the technology they are helping users with. It is crucial, however, that help desk personnel know when new systems and releases are being implemented and when users are being trained for new systems. Help desk personnel themselves should be well-trained on new systems. One sure recipe for disaster is to train users on new systems but not train the help desk personnel these same users will turn to for their support needs.

EVALUATION AND MAINTENANCE

No computerized system once implemented remains unaltered through the rest of its working life. Changes are made, and maintenance is undertaken. Hardware maintenance is usually carried out under a maintenance contract with the equipment suppliers. Hardware maintenance involves technical tasks often requiring circuitry and other specialized parts.

Software maintenance is also carried out. In time, bugs may be discovered in the programs. It is necessary to correct these. Modular program design, structured code and the documentation techniques of structured systems analysis and design ensure that this task can be effected efficiently. Software maintenance is no longer the complex and costly contribution to the entire systems project that it used to be.

User needs also evolve over time to respond to the changing business environment. Software is amended to serve these. New applications programs must be written. It is the responsibility of the computer centre or information centre to ensure that the system is kept up to date.

Some time after the system has settled down it is customary to carry out a post-implementation audit. The purpose of this audit is to compare the new system as it actually is with what was intended in its specification. To ensure independence this audit will be carried out by an analyst or team not involved in the original systems project.

The audit will consider a number of areas:

- The adequacy of the systems documentation that governs manual procedures and computer programs will be checked.
- The training of personnel involved in the use of the new system will be assessed.
- Attempts will be made to establish the reliability of systems output.
- Comparison of the actual costs incurred during implementation is made against the estimated costs, and significant variances investigated.
- Response times will be determined and compared with those specified.

The original purposes of the systems project will once again be considered. Does the system as delivered meet these objectives? The post-implementation audit will yield a report that will assess the system. Suggestions for improvements will be made. These may be minor and can be accommodated within the ongoing development of the project. If they are major they will be shelved until a major overhaul or replacement of the system is due.

During the course of the useful life of the system, several audits will be made. Some of these will be required by external bodies. Examples are financial audits required by the accountancy profession for accounting transaction processing systems. Other audits are internal. They will deal with matters such as efficiency, effectiveness, security and reliability.

GLOSSARY OF ACRONYMS

ACM	Association for Computing Machinery	**ET**	Estimated Time	**PTR**	Problem Tracking Report
AITP	Association of Information Technology Professionals	**EUC**	End-User Computing	**PV**	Present Value
		FORTRAN	FORmula TRANSlator	**PVF**	Pine Valley Furniture
API	Application Program Interface	**GDSS**	Group Decision Support System	**RAD**	Rapid Application Development
ASM	Association for Systems Management	**GSS**	Group Support System	**RAM**	Random Access Memory
		GUT	Graphical User Interface	**R&D**	Research and Development
ATM	Automated Teller Machine	**IBM**	International Business Machines	**RFP**	Request for Proposal
AT&T	American Telephone & Telegraph			**RFQ**	Request for Quote
		I-CASE	Integrated Computer-Aided Software Engineering	**ROI**	Return on Investment
BEC	Broadway Entertainment Company			**ROM**	Read Only Memory
		IDSS	Integrated Development Support System	**SDLC**	Systems Development Life Cycle
BPP	Baseline Project Plan	**IE**	Information Engineering	**SDM**	Systems Development Methodology
BPR	Business Process Reengineering	**IEF**	Information Engineering Facility		
				SNA	System Network Architecture
BSP	Business Systems Planning	**I/O**	Input/Output		
CASE	Computer-Aided Software Engineering	**IS**	Information System	**SOW**	Statement of Work
		ISA	Information Systems Architecture	**SPTS**	Sales Promotion Tracking System
CATS	Customer Activity Tracking System	**ISAM**	Indexed Sequential Access Method	**SQL**	Structure Query Language
CD	Compact Disk			**SSR**	System Service Request
CDP	Certified Data Processing	**ISP**	Information Systems Planning	**TPS**	Transaction Processing System
CD-ROM	Compact Disk-Read Only Memory	**IT**	Information Technology	**TQM**	Total Quality Management
COBOL	COmmon Business Oriented Language	**IU**	Indiana University	**TVM**	Time Value of Money
		JAD	Joint Application Design	**UML**	Unified Modeling Language
COCOMO	COnstruction COst MOdel	**KLOC**	Thousand Lines of Code		
CRT	Cathode Ray Tube	**LAN**	Local Area Network	**VB**	Visual Basic
C/S	Client/Server	**MIS**	Management Information System	**VBA**	Visual Basic for Applications
CUA	Common User Access				
DB2	Data Base 2	**M:N**	Many-to-Many	**VSAM**	Virtual Sequential Access Method
DBMS	Database Management System	**MRP**	Material Requirements Planning		
				WIP	Work in Process
DCR	Design Change Request	**MTBF**	Mean Time Between Failures	**1:1**	One-to-One
DFD	Data Flow Diagram			**1:M**	One-to-Many
DPMA	Data Processing Managers Association	**MTTR**	Mean Time to Repair Defect	**1NF**	First Normal Form
		NPV	Net Present Value	**2NF**	Second Normal Form
DSS	Decision Support System	**OA**	Office Automation	**3NF**	Third Normal Form
EDS	Electronic Data Systems	**OO**	Object-Oriented	**4GL**	Fourth-Generation Language
EFT	Electronic Funds Transfer	**OOAD**	Object-Oriented Analysis and Design		
EIS	Executive Information System	**PC**	Personal Computer		
EPSS	Electronic Performance Support System	**PD**	Participatory Design		
		PERT	Project Evaluation and Review Technique		
E-R	Entity-Relationship				
ERD	Entity-Relationship Diagram	**PIP**	Project Initiation and Planning		
ES	Expert System				
ESS	Executive Support System	**POS**	Point-of-Sale		

REFERENCES

Alavi, M., and I. R. Weiss. 1985. "Managing the Risks Associated with End-User Computing." *Journal of MIS* 2 (Winter): 5–20.

Andrews. D. C., and N. S. Leventhal. 1993. *Fusion: Integrating IE, CASE, JAD: A Handbook for Reengineering the Systems Organization.* Englewood Cliffs, NJ: Prentice-Hall.

Applegate, L. M., and R. Montealegre. 1991. "Eastman Kodak Company: Managing Information Systems Through Strategic Alliances." Harvard Business School case 9-192-030. Cambridge, MA: President and Fellows of Harvard College.

Bell, P., and C. Evans. 1989. *Mastering Documentation.* New York, NY. John Wiley & Sons.

Bohm, C., and I. Jacopini. 1966. "Flow Diagrams, Turing Machines, and Languages with Only Two Formation Rules." *Communications of the ACM* 9 (May): 366–71.

Bozman, J.S. 1994. "United to Simplify Denver's Troubled Baggage Project." *ComputerWorld,* 10(10).

Carmel, E. 1991. *Supporting Joint Application Development with Electronic Meeting Systems: A Field Study.* Unpublished doctoral dissertation, University of Arizona.

Carmel, E., J. F. George, and J. F. Nunamaker, Jr. 1992. "Supporting Joint Application Development (JAD) with Electronic Meeting Systems: A Field Study." *Proceedings of the Thirteenth International Conference on Information Systems.* Dallas, TX, December: 223–32.

Chapin, N. 1987. "The Job of Software Maintenance." *Proceedings of the Conference on Software Maintenance.* Washington DC: IEEE Computer Society Press: 4-12.

Cole, K., O. Fischer, and P. Saltzman. 1997. "Just-in-Time Knowledge Delivery." *Communications of the ACM* 40(7): 49–53.

Crowley, A. 1993. "The Help Desk Gains Respect." *PC Week* 10 (November 15):138.

Davenport, T. H. 1993. *Process Innovation: Reengineering Work through Information Technology.* Boston, MA: Harvard Business School Press.

de Jager, P. 1994. "Are We just Plain Lazy?" *ComputerWorld* 28 (February 21): 85.

Dillon, N. 1997. "Internet-based Training Passes Audit." *ComputerWorld* (November 3): 47–48.

Dobyns, L., and C. Crawford-Mason. 1991. *Quality or Else.* Boston, MA: Houghton-Mifflin.

Eason, K. 1988. *Information Technology and Organisational Change.* London: Taylor & Francis.

Fagan, M. E. 1986. "Advances in Software Inspections." *IEEE Transactions on Software Engineering* SE-12(7) (July): 744–51.

From, E. 1993. "Shouldering the Burden of Support." *PC Week* 10 (November 15): 122, 144.

Gasser, L. 1986. "The Integration of Computing and Routine Work." *ACM Transactions on Office Information Systems (*4 July): 205–25.

Ginzberg, M. J. 1981a. "Early Diagnosis of MIS Implementation Failure: Promising Results and Unanswered Questions." *Management Science* 27(4): 459–78.

Ginzberg, M. J. 1981b. "Key Recurrent Issues in the MIS Implementation Process." *MIS Quarterly* 5(2) (June): 47–59.

Hammer, M., and J. Champy. 1993. *Reengineering the Corporation.* New York, NY: Harper Business.

Hanna, M. 1992. "Using Documentation as a Life-cycle Tool." *Software Magazine* (December): 41–46.

Henderson, J. C., and M. E. Treacy. 1986. "Managing End-User Computing for Competitive Advantage." *Sloan Management Review* (Winter): 3–14.

Ives, B. and M. H. Olson. 1984. "User Involvement and MIS Success: A Review of Research." *Management Science* 30(5): 586–603.

Jones, C. 1986. *Programming Productivity.* New York, NY: McGraw-Hill.

Jones, C. 1997. "How to Measure Software Costs." *Application Development Trends* (May): 32–36.

Kling, R., and S. Iacono. 1989. "Desktop Computerization and the Organization of Work." In *Computers in the Human Context,* edited by T. Forester. Cambridge, MA: The MIT Press: 335–56.

Kling, R., and W. Scacchi. 1982. "The Web of Computing: Computer Technology as Social Organization." *Advances in Computers* 21: 1–90.

Lakhanpal, B. 1993. "Understanding the Factors Influencing the Performance of Software Development Groups: An Exploratory Group-level Analysis." *Information & Software Technology* 35(8):468–73.

Lee, D. M. S. 1986. "Usage Pattern and Sources of Assistance for Personal Computer Users." *MIS Quarterly,* 10 (December): 313–25.

Leib, J. 1997. "Baggage Suit Quietly Settled." *The Denver Post Online,* Sept. 11.

Litecky, C. R., and G. B. Davis. 1976. "A Study of Errors, Error Proneness, and Error Diagnosis in COBOL." *Communications of the ACM* 19(l): 33–37.

Lucas, H. C. 1997. *Information Technology for Management.* New York, NY: McGraw-Hill.

Lucas, M. A. 1993. "The Way of JAD." *Database Programming & Design* 6 (July): 42-49.

Markus, M. L. 1981. "Implementation Politics: Top Management Support and User Involvement." *Systems/Objectives/Solutions* 1(4): 203–15.

Martin, E. W., C. V. Brown, D. W. DeHayes, J. A. Hoffer, and W. C. Perkins, 1999. *Managing Information Technology: What Managers Need to Know.* 3rd Edition. New York, NY: MacMillan.

Martin, J. 1991. *Rapid Application Development.* New York: Macmillan Publishing Company.

Martin, J., and C. McClure. 1985. *Structured Techniques for Computing.* Englewood Cliffs, NJ: Prentice-Hall.

McConnell, S. 1996. *Rapid Development.* Redmond, WA: Microsoft Press.

Moad, J. 1993. "Inside an Outsourcing Deal." *Datamation* 39 (February 15): 20–27.

Moad, J. 1994. "After Reengineering: Taking Care of Business." *Datamation* 40 (20): 40–44.

"More Companies Are Chucking Their Computers." 1989. *Business Week* (June 19): 72–74.

Mosley, D. J. 1993. *The Handbook of MIS Application Software Testing.* Englewood Cliffs, NJ: Yourdon Press.

Nelson, R. R., and P. H. Cheney. 1987. "Training End Users: An Exploratory Study." *MIS Quarterly* 11 (December): 547–59.

Pressman, R. S. 1987. *Software Engineering: A Practitioner's Approach.* New York, NY: McGraw-Hill.

Pressman, R. S. 1992. *Software Engineering: A Practitioner's Approach.* 2d ed. New York, NY: McGraw-Hill.

Rivard, S., and S. L. Huff. 1988. "Factors of Success for End-User Computing. *Communications of the ACM* 31 (May): 552–61.

Rockart, J. F., and J. E. Short. 1989. "IT in the 1990s: Managing Organizational Interdependence." *Sloan Management Review* (Winter): 7–17.

Schneider, J. 1993. "Shouldering the Burden of Support." *PC Week 10* (November 15): 123, 129.

Schrage, M. 1993. "Unsupported Technology: A Prescription for Failure." *ComputerWorld* 27 (May 10): 31.

Schurr, A. 1993. "Support is No. I." *PC Week* 10 (November 15):126.

Tait, P., and I. Vessey. 1988. "The Effect of User Involvement on System Success: A Contingency Approach." *MIS Quarterly* 12(l) (March): 91–108.

Wood, J., and D. Silver. 1989. *Joint Application Design.* New York, NY: John Wiley & Sons.

Yourdon, E. 1989. *Managing the Structured Techniques.* 4th ed. Englewood Cliffs, NJ: Prentice-Hall.

I N D E X

CD-ROM INSTRUCTIONS

SYSTEM REQUIREMENTS

Windows PC

— 386, 486, or Pentium processor-based personal computer

— Microsoft Windows 95, Windows 98, or Windows NT 3.51 or later

— Minimum RAM: 8 MB for Windows 95 and NT

— Available space on hard disk: 8 MB for Windows 95 and NT

— 2X speed CD-ROM drive or faster

— Browser: Netscape Navigator 3.0 or higher or Internet Explorer 3.0 or higher*

— Reader: Adobe Acrobat Reader 3.0 or higher (on the enclosed CD-ROM)*

Macintosh

— Macintosh with a 68020 processor or higher, or Power Macintosh

— Apple OS version 7.0 or later

— Minimum RAM: 12 MB for Macintosh

— Available space on hard disk: 6 MB for Macintosh

— 2X speed CD-ROM drive or faster

— Browser: Netscape Navigator 3.0 or higher or Internet Explorer 3.0 or higher*

— Reader: Adobe Acrobat Reader 3.0 or higher*

* You can download any of these products using the URL below:

— **NetscapeNavigator: http://www.netscape.com/download/index.html**

— **Internet Explorer: http://www.microsoft.com/ie/download**

— **Adobe Acrobat Reader: http://www.adobe.com/proindex/acrobat/readstep.html**

GETTING STARTED

Insert the CD-ROM into your drive.

— Windows PC users should double click on My Computer, then on the CD-ROM drive. Find and double-click on the Index.html file.

— Macintosh users should double click on the CD-ROM icon on the screen, then find and double-click on the Index.html folder. (Index.html may come up automatically on the Macintosh.)

You will see an opening screen with the Welcome page and other navigation buttons. From this screen, you can click on any button to begin navigating the CD-ROM contents.

MOVING AROUND

If you have installed one of the required browsers, you will see three frames on your screen. The frame on the left-hand side contains a navigational toolbar with buttons. From this toolbar you can click on the buttons to navigate through the CD-ROM, which will then appear in the frame on the right-hand side. Note: At any time, you can use the Back button on your browser to return to the previous screen.